Music and Technology in the Twe

Music and Technology in the Twentieth Century

EDITED BY

Hans-Joachim Braun

The Johns Hopkins University Press

BALTIMORE AND LONDON

Originally published as *'I Sing the Body Electric': Music and Technology in the 20th Century*
by Wolke Verlag, 2000

Printed in the United States of America
on acid-free paper
Johns Hopkins edition, 2002

2 4 6 8 9 7 5 3

The Johns Hopkins University Press
2715 North Charles Street
Baltimore, MD 21218-4363
www.press.jhu.edu

Library of Congress Cataloging-in-Publication Data

Music and technology in the twentieth century /
edited by Hans-Joachim Braun.
p. cm.
"Technology and music at the 23rd Symposium of the International committee for the History of Technology (ICOHTEC) in Budapest, August 1996"—Pref.
Originally published: [Hofheim] : Wolke, c2000, under title:
'I sing the body electric': music and technology in the 20th century.
Includes bibliographical references and index.
ISBN 0-8018-6885-8 (cloth : alk. paper)
1. Music and technology—Congresses. 2. Electronic music—History and criticism—congresses. 3. Music—20th century—History and criticism—Congresses. 4. Music—20th century—Philosophy and aesthetics—Congresses.
I. Braun, Hans-Joachim. II. International Committee for the History of Technology. III. Symposium ICOHTEC (23rd : 1996 : Budapest, Hungary)
ML 197 .I2 2002
780'.06—dc21 2001042490

A catalog record for this book is available from the British Library.

Contents

Preface 7

Hans-Joachim Braun
Introduction 9

Geoffrey Hindley
Keyboards, Crankshafts and Communication:
The Musical Mindset of Western Technology 33

Hugh Davies
Electronic Instruments: Classifications and Mechanisms 43

Tatsuya Kobayashi
'It all Began with a Broken Organ.'
The Role of Yamaha in Japan's Music Development 59

Trevor Pinch & Frank Trocco
The Social Construction of the Early
Electronic Music Synthesizer 67

Jürgen Hocker
My Soul is in the Machine - Conlon Nancarrow -
Composer for Player Piano - Precursor of Computer Music 84

Barbara Barthelmes
Music and the City 97

Hans-Joachim Braun
"Movin' On": Trains and Planes as a theme in Music 106

Karin Bijsterveld
A Servile Imitation:
Disputes about Machines in Music, 1910-1930 121

Susan Schmidt Horning
From Polka to Punk: Growth of an
Independent Recording Studio, 1934-1977 136

Alexander B. Magoun
The Origins of the 45-RPM Record at RCA Victor, 1939-1948 . 148

Andre Millard
TAPE RECORDING AND MUSIC MAKING 158

James P. Kraft
MUSICIANS AND THE SOUND REVOLUTION:
BUSINESS, LABOR, AND TECHNOLOGY IN AMERICA, 1890-1950 168

Mark Katz
AESTHETICS OUT OF EXIGENCY:
VIOLIN VIBRATO AND THE PHONOGRAPH 174

Rebecca McSwain
THE SOCIAL RECONSTRUCTION OF A REVERSE SALIENT IN
ELECTRICAL GUITAR TECHNOLOGY: NOISE, THE SOLID BODY,
AND JIMI HENDRIX 186

Helga de la Motte-Haber
SOUNDSAMPLING: AN AESTHETIC CHALLENGE 199

Martha Brech
NEW TECHNOLOGY - NEW ARTISTIC GENRES:
CHANGES IN THE CONCEPT AND AESTHETICS OF MUSIC 207

Bernd Enders
MUSICAL EDUCATION AND THE NEW MEDIA:
THE CURRENT SITUATION AND PERSPECTIVES FOR THE FUTURE 223

INDEX 239
VITAE 253

Preface

This book has its origins in a two day session on 'Technology and Music' which I organized as part of the 23rd Symposium of the 'International Committee for the History of Technology' (ICOHTEC) in Budapest, August 1996. The topic has interested me since the late 1980s. The idea was to bring scholars together from two different fields, the history of technology and musicology. It was not the intention to deal with the topic exhaustively, but to concentrate on selected aspects. Most important were two fields: musical depictions of technology and changes in the production process of music with its aesthetic implications. Reproduction can, to an extent, be included in production.

One of the shortcomings of dealing with the relationship between music and technology is departmentalization: there are musicologists on the one hand and historians of science and technology on the other. There are those who deal only with 'art music' while looking down on 'popular music', and others who like their music 'unplugged' while denouncing those who prefer it 'plugged'.

It is not surprising that these – artificial – boundaries have prevented communication between the disciplines for quite some time. During recent years, however, the situation has improved. Another session on 'Technology and Music' took place as part of the 26th ICOHTEC Symposium in Belfort/France in 1999 and conferences like those organized by the 'Institut für Neue Musik und Musikerziehung' in Darmstadt or the bi-annual 'Klangart' in Osnabrück organized by Bernd Enders and others have advanced the cause. A pleasant byproduct of the 1996 ICOHTEC Symposium in Budapest was the foundation of an ICOHTEC Jazz group. Considering the technological theme and the fact that the annual coordination of repertoire is done by e-mail, the group is appropriately called 'E-Mail Special', a variation of the swing tune 'Air Mail Special', made famous by Benny Goodman. The group is still going strong, growing, and, hopefully, improving.

I am extremely grateful to several colleagues and friends for their advice during the preparation of this book: Mark Katz, Alex Magoun, Andre Millard, Helga de la Motte-Haber, Susan Schmidt Horning, Rebecca McSwain, Trevor Pinch, Emily Thompson and Ed Todd. I would also like to thank my research assistants in Hamburg who helped me in various ways: Reinhold Bauer, Iris Boysen, Maja Downar, Klaus Fiedler and, especially, my former *Wissenschaftliche Hilfskraft* Bernd

Schabbing, from whose thorough knowledge of the field and virtuosity on the computer I have benefited much.

Thanks are also due to Peter Mischung of Wolke Publishers for pleasant cooperation. Finally, I thank my family for their understanding that I spent less time with them than I should have done. But after this. . . .

Hans-Joachim Braun

INTRODUCTION: TECHNOLOGY AND THE PRODUCTION AND REPRODUCTION OF MUSIC IN THE 20TH CENTURY.

I

Technology has always been inseparable from the development of music. But in the 20th century a rapid acceleration took place: a new 'machine music' came into existence, electronic musical instruments were developed and composers often turned into sound researchers.[1] A problematic identification of technical change with musical progress arose. 'Noise' was raised to the state of an art[2] and, in pop and rock, electroacoustics became the essence of music making and of musical aesthetics. Recording engineers assumed increasing importance and the rise of studio aesthetics had a significant impact on the expectations of listeners in the concert hall. In view of all this, it seems correct to speak of a 'technologization' of musical aesthetics in the 20th century.

As Jaques Barzun has remarked, the moment that man ceased to make music with his voice alone, art became machine-ridden.[3] Orpheus's lyre was a machine, a symphony orchestra is a proper factory of artificial sounds, and a piano is the most appalling contrivance of levers this side of the steam engine. According to Max Weber's concept of rationalization, the natural sciences brought about a process of increasing organization and administration of nature. In his view, the ways in which composers organized their material had become increasingly rationalized and controlled through the development of new techniques. In Western art music this can be seen in the rationalization of tuning systems, in musical instrument technologies, in the development of complex notational systems and polyphony which, *inter alia*, made the development of the orchestra possible.[4]

However, the development of music technology is much more than just a reflection of technological development in general. In this volume Geoffrey Hindley claims a prominent place for music technology. He argues convincingly that the West's polyphonic harmonic art music, in which a mechanistic keyboard replaces the bardic lyre, has promoted a fundamental shift in human mental and physical orientations. This change has played a large role in the evolution of industrial culture.

Social scientists and philosophers with an interest in music have suggested that technology and musical techniques, content and meaning, generally develop together dialectically.[5] In Theodor Adorno's view, technology has gradually penetrated to the heart of the work of art itself. It has ceased to be something external to its 'inner meaning': the 'interior' and 'exterior' produce each other alternatively and mutually.[6] Adorno detected an increasing 'technification' of music technique, which is not only true of musical reproduction, but of the actual production and composition of music itself. In 1936, the cultural philosopher Walter Benjamin argued that technical reproduction destroyed the 'aura' of the original work of art; yet he also saw that this process could 'democratize' art.[7] In his contribution to this volume Alexander Magoun makes clear that the relationship between technology and music is bi-directional and that more emphasis should be put on the aesthetic implications of sound reproduction technology. Magoun points out that in investigating the relationship between musical production and consumption a broadened definition of music is required.

Musicologists, historians of technology, philosophers and social scientists have analyzed the impact of technology on music making and music reception, but musicians have also chosen themes of technology in some of their compositions. As Barbara Barthelmes shows, composers have played a considerable role in developing the myth of the city, ranging from impressionists' finely grained reflections on cities to Italian Futurists' aggressive roaring of car engines in town centres.[8] At the end of 1960s the composers R. Murray Schafer and Michael Southwort, independent of one another, started to use the term 'soundscape' in reference to the city as an urban complex; it was at this time that Schafer developed his concept of 'acoustic ecology'.[9]

The Futurists raised noise to the state of an art. Introducing his concept of 'organized sound', however, the composer Edgar Varèse did not find a mere imitation of environmental sounds satisfactory.[10] Stating his intentions he wrote:

> 'I dream of instruments obedient to my thoughts which, with their contribution of unsuspected sounds, will lend themselves to the exigencies of my inner rhythm. Why, Italian Futurists, do you slavishly reproduce only what is common-place and boring in the bustle of everyday life? The Futurists believe in reproducing sounds literally. I believe in the metamorphosis of sounds into music.'[11]

Although 'machine music' seemed to have fascinated composers like George Antheil, Karin Bijsterveld makes it clear that musicians or painters like Piet Mondriaan attributed different and even opposing characteristics to machines.[12] The editor of this volume looks at several 20th century musical works dealing with aeroplanes and railway engines and shows how composers' attitudes ranged from boundless fascination to a concern for the destructive uses of technology.[13] But in the 1920s the veneration of technology clearly dominated. In this period of *Neue Sachlichkeit* or 'New Objectivity'[14] composers seem to have forgotten the horrors of the First World War, the 'War of the Engineers'. Instead they hoped that technology would generate the means to satisfy future material and cultural needs.

II

The development of technology has always had an impact on music, but at the turn of the century musicians, inventors and 'music engineers' placed particularly great hopes on it. The Italian Futurists demanded the rejection of traditional musical principles and their substitution by free tonal expression. This led to the construction of specially designed noise instruments, the 'intonarumori'. 'Electricity the liberator' became a slogan of the day. Although mechanical organs known as orchestrions had tried to imitate the instruments of a symphony orchestra before,[15] it was Thaddeus Cahill's 'Telharmonium' of 1906 which seemed to fulfill the dreams of composers like Busoni and Varèse. Dissatisfied with the state of music of their time, they had advocated a reappraisal of the whole language of music, an enrichment of the musical alphabet and the introduction of new instruments. Cahill's 'Telharmonium' or 'Dynamophone' was a gigantic electric dynamo weighing 200 tons and employing numerous specially geared shafts and inductors to produce alternating currents of different frequencies. Via a polyphonic keyboard and associated controls these signals passed to telephone receivers where they could be heard.[16]

The Telharmonium was only an experimental device and met with little practical success, but the player piano did: it lifted musical performance out of concert halls and transferred it to private homes. This became possible by treating music as information: piano rolls stored data, not sound.[17] For many years the player piano, the 'pianola', rivaled the gramophone in popularity, opening up new possibilities for composers and artists. Making the recording, the artist gained enormous control over the outcome; processing the information or recutting the holes in the piano rolls the performance even remained malleable. The music of Conlon Nancarrow, the best known player piano composer, can be performed only by playing back his piano rolls. This method of composition, an extremely laborious process, gives the composer complete freedom in conjuring up complex contrapunctual, harmonic and rhythmic combinations which no human pianist could possibly perform.[18]

From the perspective of information technology, the player piano decodes what already exists, but does not alter or process that information. Other musical instruments developed at the time not only used data to make sound, but created information of their own. Whereas holes in the piano rolls could be seen, the case with the 'Theremin', an electric musical instrument invented by the Russian physicist and musician Lev Termen, was different: by some magic process the human hand seemed to conjure sound from the air. But in electro-acoustic terms the matter loses much of its magic. The design of his 'Theremin', 'Thereminvox' or 'Aetherophone', as he also called it, was similar to heterodyne radio reception. This is based on obtaining audible frequency beats, formed by the interference of inaudible high-frequency oscillations. In his instrument Termen applied two capacitive detectors, one a vertical rod, the other a horizontal loop. The detectors controlled pitch and amplitude, respectively, by generating electrical fields which altered according to the proximity of the performer's hands.[19]

Other electric instruments developed in various countries followed conventional designs more closely: electric or electronic pianos and electronic organs, for example, designed by 'music engineers' such as Benjamin F. Miessner or Richard H. Ranger, father of the 'Rangertone Organ'.[20]

The Hammond organ developed by the American inventor Laurens Hammond was particularly successful. By the end of 1934 Hammond designed and built an instrument with 91 small tone wheel generators, rotated by means of his newly developed synchronous motor, with harmonic drawbars placed above the keyboard to permit the mixture of different tones. Compared with other electronic organs of the time Hammond's instrument had the advantage of mechanical simplicity which made it suitable for mass production.[21] Another reason why Hammond's organ met with considerable success was that Hammond's ambitions were relatively conservative. He did not intend to create a revolutionary new instrument but wanted to design an organ which was cost efficient and relatively easy to handle.[22] The Hammond organ is an example well-known in the history of technology: in order to succeed, the step between the old and the new should not be too large.

In this the hammond organ was similar to the electric guitar. The guitar's amplification began in response to guitarists' demands for their solos to be heard through the sound of big bands. Electrification facilitated an expansion of traditional guitar solo techniques like fast runs, inflection, glissando or vibrato, while allowing new techniques which resulted in new effects like sustained tone.[23] As Rebecca McSwain points out, the electric guitar also became a vehicle for the expression of a new popular music aesthetic: with the electrification and amplification of the guitar, undesirable sounds like feedback crept in.[24] But those sounds soon became desirable and even essential. Distorted sounds from overdriven amplifiers were the first electric guitar noises to be defined as music. By a process of reconceptualization a former deficiency became a virtue and the core of a new musical aesthetic. The case of the electric guitar shows that musical instruments are not 'completed' at the stage of design and manufacture but can be 'made-over' by musicians in the process of music-making.[25]

Reconceptualization was also central to the origin of 'musique concrète' in Paris, composed from altered and rearranged sounds from the environment. 'Musique concrète' applied a concept of material used in surrealism. As Helga de la Motte-Haber shows, the latter's concept of 'objet trouvé' became part of a new, different musical logic.[26]

But composers working in electronic music studios, especially Karlheinz Stockhausen in Cologne, were not content with this. By using simple oscillators to generate electro-magnetic waves which could be translated into pure sound they aimed at fabricating from scratch any sound they could imagine. In the manner of sculptors they worked directly with their material, creating the very sounds of their compositions and hearing the result immediately. During this process, they tried to apply the principle of serialism to the structure of partial tones.[27] As Vladimir Ussachevsky, one of the founding fathers of tape music and electronic music, stated in an interview:

> 'Musique concrète was fun, playful, full of spontaneity. The Cologne people didn't want to be experimental. They wanted to be sure and very systematic, methodological and everything else. They wanted to be the exact opposite from (sic!) the French, and this is why Stockhausen made very few references to the fact that he was working in a musique concrète studio and he was there and he learned certain things there – he doesn't mention that very much.'[28]

The conflict between Paris and Cologne over different musical concepts soon dissolved. In 'Gesang der Jünglinge' (1956) Stockhausen used both natural sounds from the recording of a boy's voice and electrically generated material, and in 'Kontakte' (1960) he established contacts between the sounds of live performers and music on tape. To a large extent, and at least in theory, the composers at the Cologne studio were in a position to make Varèse's dream come true: to take an electronic signal and produce a sound with defined pitch and colour, thus opening up the possibility of sound synthesis.

But there were drawbacks. Their tools were extremely cumbersome and expensive and also very limited in their possibilities. Another method was thus called for.

Already in the mid-1930s the Russian physicist Evgenij Sholpo had applied the principle of artificially synthesizing an optical phonogram to his 'variophone'. A few years later – according to his own claims in 1938 – Evgenij Murzin, an engineer and Red Army Colonel, developed his 'ANS' sound photosynthesizer, named after the Russian composer and music reformer Alexander Nikolayevich Scriabin. The ANS octave was devided into 72 steps – rather than the customary 12 – which made it possible to synthesize an almost limitless amount of timbres. By means of the ANS Murzin transformed the process of composition into drafting; music was drawn graphically with the possibility of instant reproduction.[29]

In the United States engineers worked on similar problems. In 1945 John Hanert built the 'Hanert Electrical Orchestra', an instrument which aimed at giving the composer control over the complete fabric of musical composition. Hanert based the operation of this early electronic music synthesizer on the breakdown of a tone into characteristics such as frequency, intensity, growth, duration and timbre. Although his instrument was not economically feasible, it can be regarded as the earliest complete synthesizer which allows music to be intentionally reduced to its constituent elements and then reassembled into coherent musical structures.[30] A decade later, with the RCA synthesizer, the RCA engineers Harry Olson and Herbert Belar could, at least in theory, generate any kind of sound imaginable.

But practice was different. The RCA instrument was unable to emulate the complex sounds of Olson's target instruments, clarinet and oboe. Besides, and just as relevant, the RCA synthesizer was extremely expensive and not fit for mass production. That was remedied by Robert A. Moog, who in 1966 started to produce his Moog synthesizer. The small size and flexibility of transistors opened up the new technique of voltage control.[31] Moog devised oscillators which were controlled by the amount of voltage and could then alter the volume, pitch or overtones of the sound.[32]

More than the electronic organ, the synthesizer became an instrument in its own right. Electronic circuitry allowed it to imitate traditional instruments as well as synthesize entirely new sounds. In their contribution to this volume, Trevor Pinch and Frank Trocco examine the genesis of this first successful mass-produced synthesizer. They show that Moog, unlike the synthesizer designer Donald Buchla, who had different aesthetic ideas, came from a tradition of mass production. Pinch and Trocco also analyze how relevant social groups influenced the configuration of synthesizer construction. They point to the double character of the synthesizer. Being an outgrowth of the psychedelic culture it also dramatically influenced it. The synthesizer is a striking example of the reciprocal, co-evolutionary role of a technological artefact, embedded in a milieu which it helped to create.

From Moog to Yamaha is only a small step. Tatsuya Kobayashi interprets Yamaha's role in the development of musical instruments as one of creative adaptation, in which ingenious marketing methods and an emphasis on music education played an important role.[33] In a different context, Bernd Enders' article makes clear that in modern music education musical media competence is urgently required. New goals for music education will have to be established in order to meet future challenges.

The voltage-controlled synthesizers of Moog and Yamaha did not reach the goal Varèse and others had set. Although their electronic instruments produced a large variety of sounds their timbral capabilities remained limited. The computers and the 'digital revolution' of the 1980s helped remedy the situation. As early as 1957, Max Mathews had produced the first computer-generated sounds at the Bell Telephone Laboratories in Murray Hill, New Jersey. Bell researchers became interested in transmitting telephone information in digitized form. This was possible by first converting analogue information into corresponding numerical samples and then reversing the process. Because of the complexity involved, the computer proved immensely helpful as a development aid. Mathews, also a keen violinist, soon realized that the computer could assist composers. Composers could generate data structures from higher orders of compositional specifications. Already in 1956, Lejaren Hiller and Leonard Isaacson had tried this at the University of Illinois, experimenting with various musical styles and using calculation procedures to generate conventional musical scores. In the late 1960s computer application became more ambitious. In compositions like 'Funktion Grün', 'Funktion Orange' or 'Funktion Gelb', the German composer Gottfried Michael Koenig applied mathematically determined procedures derived from various composing programmes.[34]

The application of computers in music seemed to have no limits – every sound that could be imagined and heard could be made. Digital synthesizers treat sounds the way computers manipulate symbols. The first experiments in digital synthesis came in the late 1970s. The 'synclavier', invented in 1977, constructed every sequence of numbers, every sound, from scratch. The sampling techniques of the 1980s made it possible to treat all sound as data: once sampled, anything could be reproduced and reshaped. In 1983, the establishment of MIDI, the Musical Instrument Digital Interface, made it a simple matter to transfer digital information between various electronic instruments as well as between instruments and computers.[35]

During the last twenty years, a large number of different programmes and computer languages have been developed. Today they offer sophisticated possibilities to analyze, process and create sound in real-time. Recent technologies have made it possible to subject almost all parameters of music material, of its interpretation and, probably in the near future also of its distribution, to complete 'editability'.[36]

III

Is all this then another 'success story' of technology opening up new unlimited possibilities? Starting with MIDI, not everything is as perfect as some electronic musical instrument producers want us to believe. MIDI, a compromise between cost, performance, market preference and other factors, has many critics.[37] Artists often complain about its limited creative possibilities. Advancing the tendency toward uniformity the degree of instrument compatibility required by MIDI has resulted in a horizontal integration of the synthesizer market.[38] Using the computer as a mere compositional tool should not impair, but rather support artistic creativity in the traditional sense; the matter is, however, more complicated. Martha Brech refers to widespread apprehensions that a given computer programme or technology may have a determining role in shaping a composition, thus relegating musical inspiration to a minor rank.[39]

Such reservations are not unfounded and have a long history. Much depends on the way artistic creativity is defined. Can machines be 'creative'? The application of machines to the arts has often met with skepticism; artists and critics fear that human creativity had to be traded in for soulless schematism.[40] In the 1950s conservative musicologists, referring to electronic music, cried out against the 'dead sounds' from the loudspeaker, afraid of being swept over by those frightening manifestations of materialist culture.[41] Today, some pop musicians apply an electronic 'humanizer' to make the beats from their drum machines more irregular.

In our day, too, computers are often regarded as fettering the creative mind. Critics maintain that computer music is based in an ontology of technique instead of an ontology of music.[42] According to them the logic of musical composition is miles apart from the logic of mathematical algorithms in computers.[43] Many concertgoers, who do not have those reservations, feel a distinct uneasiness when being confronted only with loudspeakers; as a remedy they sometimes call for live-electronics, combining a live performer with electronic devices. But this attempt at combining the 'best of both worlds' often imposes new artistic limitations.[44]

The assumption of a synchronization between technological and musical process is not tenable. Historians of technology have clearly shown the ideological nature of 'progress talk' in technology[45] and changes in the aesthetics of music do not always follow innovations in music technology closely.[46] The saxophone, for example, invented about 150 years ago, played an extremely limited role in the music of the 19th and early 20th centuries. It really only 'arrived' in the late 1920s, when jazz musicians' new ideas of musical expression made the saxophone indispensable in jazz and dance music. In assigning a new role to the electric guitar, Jimi Hendrix did not so much apply recent technology, but adapted and shaped it according to his own aesthetic vision. And in using the gramophone as an instrument of musical production instead of reproduction from the late 1930s onwards, John Cage endowed it with a new role in music. Something similar goes for the pioneers of *musique concrète* from the late 1940s onwards, using wire and tape recorders in a new, creative way, or for

Karlheinz Stockhausen, who in the early 1950s found his compositional tools like sine wave generators or white noise generators in broadcasting studios, where they had been used for completely different purposes. In another context the disk jockeys of the late 1970s rose from mere record player operators to inventive musicians, producing, instead of just reproducing music.

As Barbara Barthelmes points out, it is the playful, undogmatic use of old 'objets trouvé' in a different, high tech environment, which opens up imaginative avenues of music making.[47] In the sense of the French anthropologist Claude Lévi-Strauss's 'bricolage', linking 'high tech' with 'low tech' in an unconventional way, the musician Nicolas Collins combines some musical instruments collected from flea markets with microchips; Christian Marclay uses a panel with six ancient record players to produce new, 'unheard of' music.

Musicians and engineers have not only been fascinated with the creation of new sounds, they also welcomed the possibility of generating exeptional degrees of artistic virtuosity or musical precision. Music technology could, as composers like Igor Stravinsky hoped, make 'authentic interpretation' possible. Composers would then no longer have to complain about limited insights of conductors who did not really understand their work and were free to deform it at will.[48] Who protected his work when the composer was dead and could no longer raise his voice in protest? Why should not he conduct his work himself and put it on a record as an authentic recording?

But what really is 'authentic'? Even composers who conduct their own symphonic music aren't always consistent from performance to performance; besides, they might simply be poor conductors. Even in electronic music, where everything is on tape and authenticity would seem guaranteed, there are vagaries, too: the type of tape recorder, the size of the room in which the work is being performed, the size of the audience, the positioning of the loudspeakers, all yield varying results.[49]

Although many musicians in popular music do not rate electronics very highly, most of them would probably have to find new jobs if synthesizers and samplers were abolished. But what about composers and performers in 'art music'? Looking at the last two decades their responses to technical innovations like digital recording or the use of compositional algorithms in the production of music range from an attitude of total acceptance to complete rejection.[50] There are artists with a pioneer mentality who enthusiastically welcome the possibility of being able to realize their dreams. Then there are 'reluctant pioneers' who adopt the new technology selectively and sometimes, as in 'live electronics', mix the old and the new.

On the other side of the spectrum are composers like Morton Feldman who clearly loath electronic music. To them the stereotypical DX 7 synthesizer sound is a horror. Others, who steer clear of high tech electronics, fear that they might be too old to handle the new technology professionally or that they, financially and otherwise, would find it difficult to keep up with the state of the art in music electronics. Some musicians worry that their expertise in composing or performing music might become worthless, if any technically knowledgable person could achieve similar results with little training or musical creativity.

Such fears have a long history and always arise when technological devices threaten to replace human labour. But how has technology affected the musician's work? Even in the days of Cahill's Telharmonium concert organizers not only dreamt of the 'liberation of music' by technology, but also of 'liberating' musicians from their jobs. Like capitalist entrepreneurs in other fields, they did not want to rely any longer on difficult and expensive orchestra players who might be moody or were sometimes unwilling to rehearse properly.[51] In the 1920s there was a widespread opinion among composers and music critics that the days of the large orchestras were over.[52] If not a music machine à la Cahill, then the radio and sound movies would lead to their demise. James Kraft shows how, in the United States, high-fidelity records, network radio and sound motion pictures ended a 'golden age' which musicians had enjoyed earlier this century.[53] This resulted in many lost jobs and fierce struggles against nationally oriented 'entertainment factories'. On the other hand, electrical amplification, particularly the development of the electric guitar in the 1930s, gave American musicians the freedom to travel and perform before large audiences without the capital expense of a big band.[54]

Considering the situation today, many musicians in the traditional sense of the word have become redundant. But some new jobs have opened up which do not require much traditional musical expertise: synclavier operators or keyboard players who often replace pianists and have now assumed the function of 'sound keepers'.[55]

IV

'Music engineers' like the above mentioned Lev Termen, Benjamin F. Miessner or Richard Ranger raise the issue of the relationship between music technology and military technology. This is a largely unexplored subject.[56] In the 1920s and 1930s several physicists and electrical engineers pursued research on musical instruments but also on military devices. Shortly after the First World War Termen, who was also a cellist, developed his 'Theremin' at the same time as his 'radio watchman', a capacitance alarm system which produced a whistle over headphones when anyone entered the area under surveillance. Security forces and the military jumped on this idea as well as on his 'television' device which he developed for surveillance purposes in the 1920s. His electronic listening device 'Buran', a directional microphone which in the late 1940s made it possible to eavesdrop on conversations, and his notorious 'bug' are in the same category. Detection of the latter in the U.S. embassy in Moscow led to a diplomatic *éclat* between the United States and the Soviet Union.[57]

Although not as well-known as Termen, the American inventor Benjamin F. Miessner was of the same ilk. In the 1930s he developed an electric piano and numerous electric musical instruments. Shortly after the First World War he had designed a wheeled box, his 'electric dog': a strong light pointing at the 'dog' affected photoelectric cells and operated relays which caused the dog to move towards the light. Together with the research of other engineers Miessners electric dog led to the development of burglar alarm systems similar to Termen's as well as of heat seeking missiles of the 1960s and after.[58]

Richard H. Ranger, who in the 1930s showed his electronic 'Rangertone Organ' at exhibitions together with Termen's and Miessner's instruments, can be put into a similar category. At the end of the Second World War Ranger, as head of the electronics section of the U.S. intelligence mission in Europe, transferred the technology for magnetic tape recording from Germany to the United States. This technology was used for military and civilian purposes and also for music recordings.[59]

Several researchers engaged in both music technology and military technology were associated with the Bell Telephone Laboratories in New Jersey. In the 1930s Homer Dudley developed the 'vocoder', a speech analyzer and synthesizer. This device became indispensible for speech transmission research but also for early electronic music, military communication 'encoded speech' and speech analysis for intelligence purposes.[60] In the 1960s, his colleague Winston E. Kock, a former electronic organ designer, became the first NASA chief of electronics. During the Second World War he had worked on fire control radar, later on sonar and underwater torpedo and missile control systems and eventually he became head of the National Industrial Security Association's Submarine Warfare Committee.[61] In the late 1930s Donald B. Parkinson, another researcher at Bell, worked on a music synthesizer and later developed an electrical gun coupled with an analogue computer. It was to a large extent due to this anti-aircraft system that Britain won the 'Second Battle of Britain', during which nine out of ten V-1 rockets destined for London were shot down in the late summer of 1944.[62] Finally the composer George Antheil, in cooperation with the Hollywood actress Hedy Lamarr, patented a novel radio control system for torpedos in 1942. It operated with 88 rapidly changing frequencies modelled on the way a player piano worked. After modification it became the basis for the U.S. Milstar defense system.[63]

These examples shows different applications of the same or similar physical and technical principles and that 20th century music technology and military technology often had similar origins. This does not mean that military interests were lurking behind all this. A closer look at the motivations of music engineers like Termen, Miessner or Ranger reveals that the challenge to contribute to music by providing new technological devices should not be underrated as an incentive.

V

Recording is, of course, of great interest for the development of music in the 20th century. Regarding fidelity, a weak link in the chain was the microphone.[64] In the early 1910s microphones were little more than telephone mouthpieces. Because neither low notes nor overtones could be reproduced well, recorded music lost much of its original flavour.[65] As an effective alternative, the Bell Telephone Laboratories developed a condenser microphone which went into commercial production in 1922; it became widely used in radio and, from 1925 onwards, in disk recording, bringing about a reduction in distortion and a smoother bass response.[66]

Sir Edward Elgar conducting 'The Symphony Orchestra', 10 January 1914 (P. Copeland, *Sound Recording*, London, 1991, 14.)

From the late 1920s onwards electrical recording equipment and techniques adapted from radio broadcasting offered a noticeable improvement in the frequency range of the reproduced sound. Still, with a frequency range from 100 to 5.000 cycles, musical recordings were deficient in bass and treble, but they were sensitive enough to pick up the ambience of the recording room.

In the early days of microphone popular singers applied 'crooning', a singing technique for coping with the limitations of early recording. Instead of hitting notes squarely, crooners would slide up to them, creating a sensual, undulating tone. In crooning popular singers treated the microphone as an instrument in its own right. In its public expressions of private feelings, crooning fit perfectly with the intimacy and privacy of home music listening.[67]

The basic technology of electrical recording had been developed during the First World War and derived from wireless telegraphy and the discovery of the thermionic valve. But in the United States the impetus for introducing the electrical microphone into sound recording came from declining market shares in competition with radio, which offered 'free' entertainment and better sound quality.[68]

In the late 1920s and early 1930s the relationship between the radio, recording technology and musical aesthetics was intensely debated. The composer Arnold Schoenberg remarked that radio – and the gramophone – exhibited such clear sonorities that less heavily instrumental pieces were called for.[69] Radio music of the 1920s favoured brass over strings, motoric rhythms and an anti-romantic stance,[70] all features which corresponded with the technological possibilities of the time. The German writer Bertolt Brecht aimed at the democratization of the opera and therefore advocated the new type of 'radio opera' accessible to the masses.[71] Those efforts did not meet with much success. Critics like Theodor Adorno viewed the broadcasting of music with suspicion; according to him it always ran the risk of becoming a mere potpourri, a continuous medley in the manner of a collage.[72]

As was hinted at above, avant-garde composers like John Cage used the radio and the record player as an instrument of sound generation instead of reception. In 1939 he composed his 'Imaginary Landscape No.1', a piece for electrical reproduction apparatus, in which musicians performed with recordings on variable-speed turntables. Similar compositions followed.

In Europe, particularly in Germany in the 1930s, radio stations favoured concert-hall type studios. Recording techniques aimed at lively acoustics, increasing the distance between the sound source and microphones. From the late 1930s onwards recording engineers generally used dynamic and ribbon microphones; their comparatively modest price probably encouraged their liberal use. It was the large number of microphones and mixing channels that made the most important difference between the 'dynamics' and the 'condenser' world during the 1940s and early 1950s. From the mid-1950s and early 1960s the demand for new channels grew rapidly. Sound engineers refined the methods of reverberation, multi-tracking came into common use and stereo recording offered new possibilities,[73] in particular improving spatial realism. With digital technology from the early 1980s onwards the concert hall experience again became the model for 'classical' music recordings.[74]

From the early 20th century onwards record companies and many music enthusiasts praised the blessings of audio recording. But composers and conductors showed a mixed attitude. As listeners' responses to tone tests bear out, music enthusiasts had few complaints about limited fidelity.[75] In the early days of his career, the

German conductor Wilhelm Furtwängler declined to make recordings, referring to their poor quality. Although regarding technology as a means of creating new possibilities for music, the composer Ferruccio Busoni, who was also one of the greatest piano virtuosi of the day, complained about the strain and artificiality of recording. In 1919 he wrote:

> 'Not letting oneself go for fears of inaccuracies and being conscious the whole time that every note was going to be there for eternity, how can there be any question of inspiration, freedom, swing or poetry?'[76]

But other musicians wanted to record. Conductors Leopold Stokowski and Herbert von Karajan were both fascinated by the possibilities of audio technology. From 1930 onwards, Stokowski cooperated with the Bell Telephone Laboratories in developing new methods of music reproduction.[77] In the 1960s, the pianist Glenn Gould put the recording studio at the centre of music making, relegating the live performance to the fringe. He maintained that the listener with his playback equipment could even assume the role of a recording technician, tailoring the recording to suit his individual taste and mood.[78]

Genres other than 'classical music' depended even more on recording. How could have jazz developed without it? Early on, when it was difficult to record the subtleties of a performance, jazz musicians played much like boogie-woogie players: loud, fast and somewhat one-dimensionally. Yet recording was essential for the development of jazz: recordings captured improvisations which could not – or only to a limited degree – be written down. The fact that the early recordings had only a playback time of about three minutes (three and a half at the most) had obvious disadvantages, but did make for concentration.[79] It would be interesting to place this phenomenon into the context of the technical efficiency and rationalization movement of the 1920s and 1930s.

Recording influenced music making. The sharp-witted but somewhat one-sided Theodor Adorno called Arturo Toscanini's approach to conducting 'the barbarism of perfection' and maintained that every performance of the strong-willed maestro sounded like its own phonograph record: 'machine-like' and 'perfect' or, to put it differently and exaggerating somewhat, dead, instead of alive, an example of the 'Modern Times' à la Charlie Chaplin. Ironically, however, Toscanini's style, if related to the recording technology of his time, was really a reflection of technology's imperfection: because technology was so acoustically limited with a tendency to blur and thicken the orchestral texture, the conductor's incisive and lean sound made the best of the prevailing circumstances.[80]

Making a virtue out of a *prima facie* shortcoming is a strategy which also proved successful in other instances of the technology and music relationship. The need to be economical, to say as much as possible within a short time, was already mentioned in the context of jazz improvisation. But there are other examples: as Mark Katz shows, recording has deeply influenced the aesthetics of violin playing.[81] Before the rise of sound recording, violinists generally used vibrato sparingly. But in sound recording vibrato adopted a compensatory role. It allowed violinists to overcome the limitations and liabilities of early recording equipment and also served to mask imperfect intonation – after all, recordings were made for eternity. But vibrato helped project a greater sense of the artist's presence, trying to convey another deficit of the recording studio compared to the concert hall: they made up to unseeing listeners what the musician's body language and facial expression would have communicated in a live setting. Eventually, a strong vibrato came to be accepted, and what was once considered inartistic became an aesthetic ideal.

Even in the 1950s and 1960s, when recording technology was much superior to the 1920s or 1930s, there were still skeptics like Sergiu Celebidache or Otto Klemperer. Celebidache loathed the disc, calling it a 'sound-emitting pancake', but the critical Klemperer had to realize that his worldwide fame as a conductor was to a large extent due to the existence of recording.

But what, in fact, is 'recorded'? Looking at it more closely the word 'to record' is misleading: only 'proper' live recordings, which are rare, 'record' an event, studio recordings, patched together from various 'takes', document no single occurence.[82] As to the ideas of fidelity and realism in sound recording, advances in the physical reproduction of sound have made it clear that such concepts are in fact chimeras. Indeed, improved sound reproduction technology has rather increased the difference between sound recording and sound reproduction than diminished it.[83]

Not only have musicians adapted to the conditions of sound recording but sound recordings have also influenced music listeners to such an extent that many come to the concert hall with aural expectations modelled on their experience of recorded music.[84] It seems that audiences have identified with recording technology without being aware of it. They have often come to expect a kind of dramatic, engineered sound which they know from recordings of the sixties and later. Recording technology has therefore created new aesthetic expectations which in turn have generated new artistic conventions.[85]

But what about 'live recordings'? Contrary to its name, a live recording is not a documentary photograph of a concert. As a rule, producers, in co-operation with artists and technicians, produce some artificially created 'superreality' made up from various 'live takes' which, as the Austrian pianist Alfred Brendel put it, makes the 'live recording' a link between the concert hall and the studio.[86] But even if the live recording is truly 'live', an aural image of a real concert, it still differs from the original event. Apart from recording limitations in general, music is more than 'meets the ear', it includes the 'aura' of the performance, the rapport between artist and audience, and the overall sound impression in the concert hall.[87] And if the music has strong emotional impact on the audience, musicians' errors, which would be edited out in a disc, are hardly perceived.

VI

The role of the sound engineer has changed considerably over time, demanding more and more artistic requirements. The degree to which sound engineers have taken part in aesthetic decision-making has undoubtely increased. This is particularly true of popular music, but also of 'art music' and has led to significant changes in its aesthetics.[88]

Because performances were recorded directly on disc or single track tape, little editing was possible until late in the 1940s. The prevailing aesthetics of 'concert hall realism' and 'high fidelity' aimed at a recording which captured the sound of a performance as closely as possible. The recording companies and broadcasting corporations built studios which tried to simulate the psychoacoustics of a live performance. In Germany, sound engineering became an academic profession when the Music Institute at Detmold introduced a course in 1946;[89] two years later the Audio Engineering Society established audio engineering as a separate profession in the United States.[90]

Especially in popular music, tape recording of the late 1940s started to undermine the predominance of the former craft-union mode of studio collaboration and made the decentralization and democratization of recording possible. A new 'entrepreneurial mode' of collaboration arose in which entrepreneurs, independent studio owners and sound engineers with only small resources brought about a shift in music aesthetics. They no longer sought to emulate a live performance, but regarded studio recording as aesthetically desirable in itself. To achieve this, they applied various electro-acoustic tricks: echo and reverberation devices, equalizers and filters, new microphone placements and different forms of tape editing. In recordings of 'classical music' they tried to obtain a seemingly 'natural' concert hall sound by applying 'artificial' editing processes.[91]

The entrepreneurial mode in pop music allowed an easier exchange of ideas and skills among musicians, technicians and music producers. Sound engineers not only shared control of studio technology with musicians, but a growing technical awareness of the artists moved the balance of power from engineers and record producers to the performers themselves.[92]

During the 1960s the development of recording technology facilitated the pop musician's involvement in sound engineering. As Susan Schmidt Horning points out, the spread of affordable tape recorders gave small, independent recording studios an opportunity to challenge the hegemony of major labels. Forming the nexus between the emerging fields of audio engineering and popular music, the independent studio became a site of electro-acoustic invention, making an impact on musical performance and composition.

In the 1960s the tape recorder became central to the production of popular music as the recording studio, rather than the concert hall, became the predominant venue for music making. At least in pop music, Glenn Gould's predictions have proved correct. Rock musicians regarded studio equipment, once considered 'merely machines', as essential musical instruments. Digital musical instruments and recording devices were no longer separate technologies: with the rise of synthesizers, samplers and sequencers during the late 1970s and early 1980s the trend in popular music toward the fusion of musical instruments and recording devices seems to have become complete. The distinction between music production and consumption has become blurred and almost meaningless.[93]

Andre Millard investigates how the cassette recorder and the ubiquitous compact cassette tape have become a major force not only in the dissemination, but in the creation of rap in the 1980s and of industrial and house music in the 1990s. In rap, the technology has completely taken over the process of making the music instead of just recording and reproducing it.[94] In 'Karaoke', in which music listeners – consumers – are invited to sing along with their favourite songs and, performing with pre-recorded arrangements of popular hits, take on the role of lead vocalist, there is 'participatory consumption': production within consumption.[95]

In both popular and 'art music' several sound engineers gained prominence. No longer relegated to a place behind the scene they developed different sound engineering philosophies and have acquired the status of stars. Engineers became associated with a particular type of 'sound' featured by a label. Atlantic's Tom Dowd managed to produce a striking clarity of sound, which contrasted with the 'RCA mix', in which just one or two microphones were used.[96] In Jazz there is – *inter alia* – the highly regarded Rudy van Gelder, who started with a studio in his living room. From 1953, labels like Blue Note, Prestige, and Savoy favoured his services because of his spacious, but at the same time clear and detailed sound.[97] In classical recording, a producer like Columbia's Andrew Kazdin used closer miking than most of the Europeans, spotlighting solo instruments.[98] Kazdin sacrified ambiance for analytic clarity and an almost tactile proximity. In his 'sonic stage' approach, the Decca engineer John Culshaw developed a novel aesthetic definition of musical recordings. Culshaw felt that modern recording technology should be used freely to create an ideal 'sound stage'. This approach allowed him to produce aurally dramatic effects which were called for in the opera score, but could not be realized during the actual performance.[99]

In view of the sophisticated relationship between recording technology and musical aesthetics, it should be no surprise that differences of opinion between conductors, producers and sound engineers occurred. One might think of prestigious EMI producers like Fred Gaisberg, Walter Legge or Suvi Raj Grubb and their relationship with various prominent conductors. Otto Klemperer's shocked exclamation, 'Lotte, ein Schwindel', is well known: producer Grubb had corrected a minor mistake in an otherwise exemplary take through editing which made Klemperer complain to his daughter about such technical trickery.[100] It is quite possible that the conductor thought the wrong note would not be heard. He was probably listening with the ears of a live performer, whereas the producer and recording engineer listened with technologically conditioned ears.[101] But during the last three decades or so a *rapprochement* has taken place. Sound engineers are required to know a lot about music and

Whilst recording Donizetti's *Lucia di Lammermoor* in the Kingsway Hall, London, in 1959, the producer Walter Legge and the conductor Tullio Serafin realize that something is not right.

"We have a problem", muses Legge as Serafin wonders what to do.

"This is how it must be done", Legge demonstrates to a bemused Serafin.

Tullio Serafin (conductor) and Walter Legge (producer) recording Donizetti's 'Lucia di Lamamoor' in Kingsway Hall, London 1959. (P. Martland, *Since Records Began. EMI. The First Hundred Years* [London, 1997], 167.)

many conductors have become well-versed in matters of audio technology.

VII

Compared with the beginning of the 20th century, reservations against the reproduction of music by audio media have nearly vanished today. This has to do with significant improvements in audio engineering, but also with the general acceptance of electrification in all spheres of life.[102]

But a revolution in music making as a consequence of electrification and electronics, as music engineers of the 1920s and 1930s and electronic music composers like Stockhausen in the 1950s had predicted,[103] has taken place only in a few music genres. In 'art music', music production has proved remarkably resistant to the aims of electronic and computer music enthusiasts. Instead of trying to come to grips with this genre, some performers have concentrated their energy on problematic endeavours like 'authentically' interpreting 'classical' music with recently manufactured 'period instruments'.[104] Composers of 'minimal music' often feel disturbed by electronic sounds and prefer their music 'unplugged'.

In most fields of popular music 'unplugging' can hardly be imagined, because electricity is the life and soul of the whole undertaking. If unplugged, not much would be left. In jazz, however, most musicians do without electrification and electronic means, although they do use 'obligatory' microphones and amplify their guitars.

In 'art music' there seems to be a great potential in exploring and developing new artistic genres like sound performance, sound sculptures and sound installations, combining audio and visual elements. Martha Brech surveys of them in her contribution to this volume. Those genres will, hopefully, reach a wider audience than at present, leaving the seclusion of the avant-garde behind them and becoming more 'popular'. With 'virtual reality' and 'interaction' (for whatever the latter is worth at present) a cultural ambiance congenial to them certainly exists.

This points to a related topic which Helga de la Motte-Haber explores in her article. In our age of the mediated perception of the environment it seems increasingly difficult to come to grips with 'reality'. By pushing a button we can order a sunset whenever we feel like it. It is therefore incumbent on today's artists to carry through experiments in perception, reflect on the conditions of perception and invite the perceiving subject to become part of them. Compared with the sublime aims of art in the romantic era this goal might be modest. But it is certainly important.

Notes

1 For recent book length surveys see J. Stange, *Die Bedeutung der elektroakustischen Medien für die Musik des 20. Jahrhunderts* (Pfaffenweiler, 1989); P. Manning, *Electronic and Computer Music*, 2nd ed., (Oxford, 1993); J. Chadabe, Electronic Sound: *The Past and Promise of Electronic Music* (Upper Saddle River, N.J., 1997); M. Supper, *Elektroakustische Musik und Computermusik. Geschichte – Ästhetik – Methoden – Systeme* (Hofheim, 1997), and A. Ruschkowski, *Elektronische Klänge und musikalische Entdeckungen* (Stuttgart, 1998). On listening see K.-H. Blomann and F. Sielecki (eds.), *Hören – Eine vernachlässigte Kunst?* (Hofheim, 1997). There are several collections of articles on the relationship between technology and music, based upon recent conferences on the theme. See, for example, B. Enders and S. Hanheide (eds.), *Neue Musiktechnologie* (Mainz, 1993); B. Enders, *Neue Musiktechnologie*, vol. 2 (Mainz, 1996); H. de la Motte-Haber and R. Frisius (eds.), *Musik und Technik* (Mainz, 1996), B. Enders and N. Knolle (eds.), *KlangArt-Kongreß 1995* (Osnabrück, 1998).

2 On music and noise see e.g. R. M. Schafer, *The Tuning of the World* (Toronto, New York, 1977); K. Hübner, *Lärmreise. Über musikalische Geräusche und geräuschvolle Musik* (Augsburg, 1992); D. Toop, *Ocean of Sound: Aether Talk, Ambient Sound and Imaginary Worlds* (Baltimore, 1996); R. Liedtke, *Die Vertreibung der Stille* (Munich, 1996); J. Attali, *Noise. The Political Economy of Music* (Manchester, 1977); S. Sanio and C. Scheib (eds.), *Das Rauschen, Aufsätze zu einem Themenschwerpunkt im Rahmen des Festivals »Musikprotokoll« 95 im steirischen Herbst* (Hofheim, 1996).

3 E. Eisenberg, *Recording Angel: Music, Records and Culture from Aristotle to Zappa* (New York, 1987), 176.

4 M. Weber, *The Rational and Social Foundations of Music* (1911), transl. and ed. by D. Martindale, J. Riedel and G. Neuwirth (Carbondale, IL, 1958); M. Paddison, *Adorno, Modernism and Mass Culture. Essays on Critical Theory and Music* (London, 1996), 23; A. Smudits, 'Technik und musikalisches Handeln. Zur Aktualität Max Webers im Zeitalter der Elektronik', in A. Smudits and H. Staubmann (eds.), *Kunst – Geschichte – Soziologie: Beiträge zur soziologischen Kunstbetrachtung in Österreich. Festschrift für Gerhart Kapner* (Frankfurt am Main, 1997), 157-171; H. Neuhoff, 'Rationalitätsgrade oder Rationalitätstypen? Zur Kritik von Max Webers »Musiksoziologie«, in R. Kopiez and others (eds.), *Musikwissenschaft zwischen Kunst, Ästhetik und Experiment. Festschrift Helga de la Motte-Haber zum 60. Geburtstag* (Würzburg, 1998), 411-418.

5 See R. Middleton, *Studying Popular Music* (Milton Keynes, 1990), 90.

6 T. Adorno, 'Music and Technique' (1958), transl. W. Blomster, *Telos*, 1977, 32: 80. See also Paddison (note 4), 102 and, by the same author, *Adorno's Aesthetics of Music* (Cambridge, 1993); O. Kolleritsch (ed.), *Adorno und die Musik* (Graz, 1979), and T. Levin, 'For the Record: Adorno on Music in the Age of its Technological Reproducibility', October (Winter 1990), 23-47.

7 W. Benjamin, 'The Work of Art in the Age of Mechanical Reproduction' (1936), in *Illuminations*, ed. H. Arendt, trans. H. Zohn (London, 1970), 217-51; G. Klein, *Electronic Vibrations. Pop Kultur Theorie* (Hamburg, 1999), 100.

8 See B. Barthelmes' article in this volume.

9 Schafer's main publication is *The Tuning of the World* (see note 2).

10 On Varèse, *inter alia*, H. de la Motte-Haber, *Die Musik von Edgar Varèse. Studien zu seinen nach 1918 entstandenen Werken* (Hofheim, 1993).

11 Letter to T. H. Greer, Midwestern University, 15 August 1965, Luening Papers, Box 46, New York Public Library, Music Division, Special Collections.

12 See K. Bijsterveld's contribution to this volume.

13 Apart from his contribution to this volume see also his 'Technik im Spiegel der Musik des frühen 20. Jahrhunderts', *Technikgeschichte* 1992, 59: 108-131 and 'Technik als Thema der Musik des 20. Jahrhunderts', *Technik und Kunst*, ed. D. Guderian (Düsseldorf, 1994), 376-404.

14 T. P. Hughes, *American Genesis. A Century of Invention and Technological Enthusiasm 1870-1970* (New York, London, 1989). On music of 'Neue Sachlichkeit' N. Grosch, *Die Musik der Neuen Sachlichkeit* (Stuttgart, Weimar, 1999).

15 H. Davies, 'A History of Sampling', *Organised Sound. An International Journal of Music Technology* 1996, 1, No. 1-2: 3-11, 4. See also Hugh Davies' contribution to this volume.

16 R. Weidenaar, *Magic Music from the Telharmonium* (Metuchen, N.J., 1995).

17 For this and the following T. Levenson, *Measure for Measure. How Music and Science together have explored the Universe* (New York, 1997).

18 See J. Hocker's contribution to this volume and also his 'Von der Klangwolke zur Tonkaskade. Akustische Möglichkeiten des Selbstspielklaviers' in *Welt auf tönernen Füßen. Die Töne und das Hören* ed. Kunst- und Ausstellungshalle der Bundesrepublik Deutschland (Göttingen, 1994), 401-410 and 'Ohne Grenzen. Musik für Player Piano', *Neue Zeitschrift für Musik* 1995, No. 2: 20-29.

19 B. M. Galeyev, 'L.S. Termen: Faustus of the Twentieth Century', *Leonardo*, 1991, 24, No. 5: 573-579, 575; Manning (note 1), 3. Also M. Lobanova, 'Erfinder, Tschekist, Spion. Das bewegte Leben des Lew Termen', *Neue Zeitschrift für Musik* 1999, No. 4: 50-53.

20 On Miessner see H. Davies, 'Benjamin F. Miessner', *The New Grove Dictionary of Musical Instruments* (New York, 1984), vol. 2 and T. L. Rhea, 'The Evolution of Electronic Musical Instruments in the United States' (Phil. Diss., George Peabody College, Nashville, TN, 1972). On Ranger see D. L. Morton jr., 'The History of Magnetic Recording in the United States, 1888-1978' (Phil. Diss., Georgia Institute of Technology, Atlanta, GA, 1995) and his *Off the Record: The Technology and Culture of Sound Recording in America* (Piscataway, NJ, 2000)

21 Rhea (note 20), 147.

22 P. Théberge, *Any Sound You Can Imagine. Making Music / Consuming Technology* (Hanover, London, 1997), 45.

23 Middleton (note 5), 90.

24 See her contribution to this volume.

25 Théberge (note 22), 160.

26 H. de la Motte-Haber's article in this volume. On the origins of musique concrète see the book of its main protagonist P. Schaeffer, *A la recherche d'une musique concrète* (Paris, 1952). There is a very informative survey of musique concrète in R. Frisius, 'Musique concrète' in *Musik in Geschichte und Gesellschaft*, vol. 6 (Kassel, Stuttgart, 1997), 1834-1844.

27 P. Griffith, *Modern Music. A Concise History* rev. ed. (New York, 1994), 147; Levenson (note 17), 296. On the origins of electronic music in Cologne see E. Ungeheuer, *Wie die elektronische Musik 'erfunden' wurde...Quellenstudie zu Werner Meyer-Epplers Entwurf zwischen 1949 und 1953* (Mainz, 1992). Also G. Borio, 'New Technology, New Techniques: The Aesthetics of Electronic Music in the 1950, *Interface, Journal of New Music Research*, 1993, 22: 77-87; and E. Ungeheuer and P. Decroupet, 'Technik und Ästhetik der elektronischen Musik' in *Musik und Technik* (note 1), 123-124. On the related aspects of sound microscopy see I. Pintér's article in this volume.

28 In interview No. 4 at the Columbia-Princeton Electronic Music Center, 14 April 1977; Ussachevsky Collection, Box 39, Library of Congress, Music Division, Washington D.C.

29 Galeyev (note 19), 577. Also F. K. Prieberg, *E. M. Versuch einer Bilanz der elekronischen Musik* (Freudenstadt, 1980), 165-170 and S. Khankaev, 'Der optoelektrische Sythesizer ANS', in Deutsche Gesellschaft für Elektroakustische Musik, *Mitteilungen* 1997, 27: 8-11.

30 Rhea (note 20), 190.

31 R. A. Moog, 'Electronic Music', *Journal of the Audio Engineering Society* 1977, 25, No. 10/ 11: 855-861; also Levenson (note 17), 299.

32 A. Millard, *America on Record. A History of Recorded Sound* (Cambridge, New York), 1995, 303-304.

33 See T. Kobayashi's contribution to this volume. See also B. Johnstone's, *We Were Burning* (New York, 1999) for Yamaha's interest in technological innovation.

34 Manning (note 1), 175-6, 218-9, 244.

35 Levenson (note 17), 302.

36 N. Knolle and A. Weidenfeld, ''Unplugged' – Stationen der Produktion, Distribution und Rezeption von Musik unter dem Eindruck von Technik' in B. Enders and N. Knolle (eds.), *KlangArt-Kongreß 1995*, (Osnabrück, 1998), 49-70, 60. On IRCAM see G. Born, *Rationalizing Culture. IRCAM, Boulez and the Institutionalization of the Musical Avant-Garde* (Berkeley, Los Angeles, London, 1995).

37 Chadabe (note 1), 185. On the origins of MIDI see R. Aicher, 'Alle Menschen werden Brüder. Die kurze Geschichte der MIDI-Schnittstelle', *Keys* 1989, Sept./ Oct.: 80-85. Also G. Loy, 'Musicians make a Standard: The MIDI Phenomenon', *Computer Music Journal* 1985, 9, No. 4: 8-26; D. Rossum, 'Digital Music Instrument Design: The Art of Composing', in *The Proceedings of the AES 5th International Conference: Music and Digital Technology* (Los Angeles, 1987), 21-25; F. R. Moore, 'The Dysfunctions of MIDI', *Computer Music Journal* 1988, 1, Nr. 1: 19-28 and A. Goodwin, 'Sample and Hold: Pop Music in the Digital Age of Reproduction', *Critical Quarterly* 1988, 30, No. 3: 34-49.

38 Loy (note 37), 20.

39 See M. Brech's article in this volume. Also W. A. Schloss and D. A. Jaffe, 'Intelligent Musical Instruments: The Future of Musical Perfomance or the Demise of the Performer', *Interface.*

Journal of New Music Research, 1993, 22: 183-193; N. Knolle, 'HUMAN OUT und MIDI IN? – Anmerkungen zur Subjektseite der Computerisierung des Musikmachens' in *Neue Musiktechnologie* (note 1), 384-401.

40 See for example the chapter 'Faszination und Schrecken der Maschine: Technik und Kunst' in H.-J. Braun and W. Kaiser, *Energiewirtschaft, Automatisierung, Information seit 1914, Propyläen Technikgeschichte*, vol. 5 (Berlin, 1992), 255-279.

41 F. Blume, *Was ist Musik* (Kassel, 1959), 17.

42 K.-P. Richter, *Soviel Musik war nie. Von Mozart zum digitalen Sound. Eine musikalische Kulturgeschichte* (Munich, 1997), 171.

43 K. Böhmer, 'Phantasie über Technologie', *Neue Zeitschrift für Musik*, 1996, No. 5: 38-43, 42. Similarly critical is J. Fritsch 'Kreativität aus dem Chip? Zwei Positionen', in Der Bundesminister für Bildung und Wissenschaft (ed.), *Neue Technik – Neue Medien – Neue Musik? Aus- und Weiterbildung – Arbeitsmarkt – Rechtsprobleme – Grundsatzfragen – Kultur und Technik* (Bonn, 1990), 69-72 whereas J. P. Fricke, *ibid.*, 72-75 sees the potential of computers in music in a much more positive light.

44 On live-electronics and 'interactive' systems see H. de la Motte-Haber, 'Ästhetische Innovationen in der elektro-akustischen Musik. Neue Klänge und Klangsynthesen: der Weg aus der Tradition', in C. Ballmer and T. Gartmann (eds.), *Tradition und Innovation in der Musik. Festschrift für Ernst Lichtenhahn zum 60. Geburtstag* (Winterthur, Switzerland, 1993), 213-223, 220-223. On live-electronics J. Fritsch, 'Live-Elektronische Musik. Musiker und Instrument im Zeitalter der Elektronik' in *Neue Musiktechnologie* vol. 2 (note 1), 67-73. On the origins of algorithmic structures in electronic music see H. de la Motte-Haber, 'Von der Maschinenmusik zur algorithmischen Struktur', in *Musik und Technik* (note 1), 79-88.

45 J. Staudenmaier's contribution in M. Roe Smith and L. Marx (eds.), *Does Technology Drive History? The Dilemma of Technological Determinism* (Cambridge, MA and London), 1994, 259-273

46 P. N. Wilson, 'Aufbruch ins Jenseits. Medientechnologie und musikalische Kreativität – ein Ensemble der Widersprüche', in *Step Across the Border. Neue musikalische Trends – neue massenmediale Kontexte.* Beiträge zur Popularmusikforschung 19/29, ed. H. Rösing (Karben, 1997), 92-102, on this and the following.

47 B. Barthelmes, 'Musikklitterung, Electronics und Klangaquarien. Klangbasteleien in der zeitgenössischen Musik', *Positionen*, 1995, 25: 11-16, and the same author's contribution to this volume.

48 K. Blaukopf, *Werktreue und Bearbeitung. Zur Soziologie der Integrität des musikalischen Kunstwerks* (Karlsruhe, 1968), 41.

49 Stockhausen realized this already in 1953 and although conditions have altered somewhat, the problem still exists. Wilson (note 46), 98.

50 For this I draw on the paper given by R. Waschka II on 'Artist Responses to Changes in Music Technology' at the KlangArt-Congress on 'Global Village, Global Brain, Global Music', Osnabrück, 10-13 June 1999. I am grateful to Rodney Waschka for permission to quote from his paper.

51 F. K. Prieberg, *Musica ex Machina. Über das Verhältnis von Technik und Musik* (Berlin, 1960), 201.

52 H. H. Stuckenschmidt, 'Die Mechanisierung der Musik', in H. H. Stuckenschmidt, *Die Musik eines halben Jahrhunderts 1925-1975. Essays und Kritik* (Munich, Zurich, 1976), 9-15.

53 See J. P. Kraft's article in this volume and also his *Stage to Studio. Musicians and the Sound Revolution 1890-1950* (Baltimore, 1996).

54 S. Frith, 'Art versus Technology: The Strange Case of Popular Music', *Media, Culture and Society* 1986, 8: 263-279, 273.

55 H.-J. Braun, 'I sing the Body Electric: Der Einfluß von Elektroakustik und Elektronik auf das Musikschaffen im 20. Jahrhundert', *Technikgeschichte* 1994, 61: 353-373. Also B. Schabbing, 'Musiktechnologie: Die Entwicklung von elektrischen und elektronischen Klangerzeugern im 20. Jahrhundert', *Muziek & Wetenschap. Dutch Journal for Musicology* 1997, 6, no. 2: 137-156. On the influence of music technology on popular music see B. Enders, 'Der Einfluß moderner Musiktechnologien auf die Produktion von Popularmusik', in M. Heuger and M. Prell (eds.) *Popmusik Yesterday, Today, Tomorrow. 9 Beiträge vom 8. Internationalen Studentischen Symposium für Musikwissenschaft in Köln 1993*, (Regensburg, 1995), 47-71.

56 For this and the following see my forthcoming article 'Strange Bedfellows'. The Relationship between Music Technology and Military Technology in the First Half of the Twentieth Century', which will appear in the collection of papers given at the 1999 KlangArt-Congress in Osnabrück (see note 50).

57 There is some information on Termen in Galeyev (note 19).

58 On Miessner see Davies and Rhea (note 20).

59 On Ranger see Morton's Ph. D. thesis (note 20).

60 On Dudley briefly: M. R. Schroeder, 'Homer W. Dudley: A Tribute', *The Journal of the Acoustical Society of America*, 1981, 69, No. 4: 1222.

61 W. E. Kock, *The Creative Engineer. The Art of Inventing* (New York, 1978).

62 'David B. Parkinson's Gun Director', *Colliers Magazine*, 1945, May 12: 24.

63 H.-J. Braun, 'Advanced Weaponary of the Stars', *American Heritage of Invention & Technology*, 1997, 12, No 4: 10-16.

64 For this and the following P. K. Burkowitz, 'Recording, Art of the Century', *Journal of the Audio Engineering Society* 1977, 25, No. 10/ 11: 873-879. Also by the same author *Die Welt des Klangs. Musik auf dem Weg vom Künstler zum Hörer* (Stuttgart, 1995) as well as W. M. Berten, *Musik und Mikrofon* (Düsseldorf, 1951).

65 S. Struthers, 'Technology in the Art of Recording' in A. L. White (ed.), *Lost in Music. Culture, Style, and the Musical Event* (London, New York, 1987) 241-258, 243.

66 T. Gioia, in his otherwise excellent *The History of Jazz* (New York, Oxford, 1997), 73, underrates the significance of early electrical recording for recording music.

67 Middleton (note 5), 85; M. Chanan, *Repeated Takes. A Short History of Recording and its Effects on Music* (London, New York, 1995), 68; Frith (note 54), 270.

68 Struthers (note 65), 252-3.

69 Chanan (note 67), 116.

70 H. de la Motte-Haber, 'Radio (Un) Kultur' in Thomas Vogel (ed.), *Über das Hören. Einem Phänomen auf der Spur*, (Tübingen, 1996), 145-157, 149.

71 M. E. Cory, 'Soundplay: The Polyphonous Tradition of German Radio Art', in D. Kahn and G. Whitehead (eds.), *Wireless Imagination. Sound, Radio, and the Avant-Garde* (Cambridge, MA, London, 1994), 331-371.

72 Chanan (note 67), 118.

73 Burkowitz (note 64), 876-7.

74 J. Stolla, 'Schallaufzeichnungstechnik und Klanggestaltung. Aufnahmetechnik, Klanggestaltung und Technikrezeption bei Aufnahmen klassischer Musik seit 1950', *Technikgeschichte* 1998, 65, No. 2: 121-140.

75 This is shown in E. Thompson, 'Machines, Music and the Quest for Fidelity: Marketing the Edison Phonograph in America, 1877-1925', *The Musical Quarterly* 1995, 79, No. 1: 131-171, esp. 156-159.

76 For a short survey on Busoni see R. Ermen, *Ferruccío Busoni* (Reinbek near Hamburg, 1996). The quotation is in Chanan (note 67), 52.

77 R. E. McGinn, 'Stokowski and the Bell Telephone Laboratories: Collaboration in the Development of High-Fidelity Sound Reproduction', *Technology and Culture* 1983, 24, No. 1: 38-75.

78 G. Gould, 'The Prospects of Recording', *High Fidelity* 1966, 16, April: 46-63 and B. Rzehulka, 'Abbild oder produktive Distanz? Versuch über ästhetische Bedingungen der Schallplatte', in M. Fischer, D. Holland and B. Rzehulka (eds.), *Gehörgänge. Zur Ästhetik der musikalischen Aufführung und ihrer technischen Reproduktion* (Munich, 1986), 85-114, 86.

79 B. Priestley, *Jazz on Record* (New York, 1991), 34-35.

80 H. Sachs, *Toscanini* (New York, 1981), also Channan (note 67), 118-122.

81 See M. Katz's contribution to this volume. See also his illuminating 'Making America More Musical through the Phonograph, 1900-1930, *American Music* 1998, 16: No.4: 448-475. Also R. Philip, *Early Recordings and Musical Style. Changing Tastes in Instrumental Performance, 1900-1950* (Cambridge, New York, 1992).

82 Eisenberg (note 3), 109.

83 Struthers (note 65), 243. Also M. Krause, 'Original oder Illusion: Ziel der technischen Reproduktion von Musik?' in R. Kopiez and others (eds.), *Musikwissenschaft zwischen Kunst, Ästhetik und Experiment. Festschrift Helga de la Motte-Haber zum 60. Geburtstag* (Würzburg, 1998), 319-329.

84 J. Badal, *Recording the Classics. Maestros, Music, and Technology*, (Kent, OH, London, 1996), 2.

85 J. Frederickson, 'Technology and Performance', *International Review of the Aesthetics and Sociology of Music* 1989, 20, No. 2: 193-220, 199.

86 Rzehulka (note 78), 86, 109.

87 H. Rösing, 'Digitale Medien und Musik – Dritte Revolution oder Fortführung des Elektrifizierungsprozesses', in *Step Across the Border. Neue musikalische Trends - neue massenmediale Kontexte*, Beiträge zur Popularmusikforschung 19/ 20, ed. H. Rösing (Karben, 1997), 104-125, 121, and by the same author 'Musik – ein audiovisuelles Medium. Über die optische Komponente der Musikwahrnehmung', in R. Kopiez and others (eds.), *Musikwissenschaft zwischen Kunst, Ästhetik und Experiment. Festschrift Helga de la Motte-Haber zum 60. Geburtstag* (Würzburg, 1998), 451-463.

88 H.-L. Feldgen, 'Anmerkungen zur Entwicklung des Tonmeisterberufes', *11. Tonmeistertagung* (Berlin, 1978), 82-88. Also H.-P. Reinecke, 'Zum Berufsbild des Tonmeisters', *8. Tonmeistertagung* (Hamburg, 1969), 57-59; W. John, 'Nochmals: Ist der Tonregisseur ein Interpret?', *Musik und Gesellschaft* 1987, No. 2: 381-383, and H. Jäckl, 'Mittler und Partner von Komponisten und Interpreten. Zur Arbeit des Tonmeisters in der medialen Musikproduktion' *Musik und Gesellschaft* 1989, No. 11: 597-599.

89 'Thienhaus, Erich', in C. Dahlhaus (ed.), *Riemann Musiklexikon*, Ergänzungsband Personenteil, (Mainz, 1975), 777. Also Stolla (note 74), 123.

90 E. R. Keely, 'From Craft to Art: The Case of Sound Mixers and Popular Music', *Sociology of Work and Occupations* 1979, 6, No. 1: 3-29.

91 Keely (note 90), 13.

92 Millard (note 32), 292.

93 Théberge (note 22), 194, 206, 21.

94 See A. Millard's contribution to this volume. Also B. Eno, 'The Studio as Compositional Tool – Part I and II', *Downbeat* 1983, 50, No. 7/8: 56-57, 50-53, respectively.

95 Théberge (note 22), 251-2.

96 Chanan (note 67), 104. On the role of the producer in popular music see A. Hennion, 'An Intermediary between Production and Consumption: The Producer of Popular Music', *Science, Technology & Human Values* 1989, 14, Nr. 4: 400-424.

97 Priestley (note 79), 114.

98 Eisenberg (note 3), 122.

99 Rzehulka (note 78), 96-98; Badal (note 84), 7. See also J. Culshaw's *Putting the Record Straight* (London, 1981).

100 S. R. Grubb, *Music Makers on Record* (London, 1986), chapter 13.

101 Frederickson (note 85), 200.

102 D. Nye, *Electrifying America: Social Meanings of a New Technology, 1880-1940* (Cambridge, MA, 1990).

103 In the early 1950s Stockhausen said to Heinz Schütz, sound engineer of the Cologne studio, that in twenty years time nobody would speak anymore of Bach and the classical composers: M. Kurtz, *Stockhausen. Eine Biographie* (Kassel, Basel, 1988), 93. On Stockhausen see especially R. Frisius, *Karlheinz Stockhausen. Einführung in das Gesamtwerk. Gespräche mit Karlheinz Stockhausen* (Mainz, 1996).

104 Richter (note 42), 161-165.

Geoffrey Hindley

KEYBOARDS, CRANKSHAFTS AND COMMUNICATION: THE MUSICAL MINDSET OF WESTERN TECHNOLOGY

Postulate: Music is the most immediate of the arts : the one most intimately related with the brain's neural patterns.
Proposition: The mindset of western art music is at the core of modern world techno-culture.

In the year 1887, Torakusa Yamaha, then aged thirty six, built a western keyboard instrument he called an 'organ'. In 1900, his company built its first piano. Since then, of course, Yamaha have diversified into many types of engineering. To a western observer, unused to motorbikes and concert grands coming from the same stable, this may seem odd. It should not, for, as I aim to show, western technology and the west's peculiar tradition of classical music are a symbiotic pair, twin children of a fissure in the human psyche which began in Europe about the year 1000. It made possible the shift to the Age of Mechanical Technology, the third great age in human development, successor to the Hunter Gatherers and the Agricultural Revolution.

I shall argue that the mechanical habit of mind, that the characteristically western pattern of progress thinking, that the habit of innovation, that even the basic principles of 'R & D', originated with the new mindset evolved in and by the western European tradition of art music.

It is no coincidence that music and technology are the two all pervasive western exports to the world at large. For centuries, the church organ – that uniquely western machine – was the largest mechanism, even including the city or cathedral clock, in any European town. It was no coincidence that Yamaha should build an organ and then rapidly diversify into engineering. It is no coincidence that Japan, the first non-western culture to adopt western style sci-tech, should also be the first to develop a tradition of western style classical music, nor that Korea has followed in its path.

Now, compared with all the other high musical tradition, western classical music is very peculiar indeed. From ancient Greece to traditional Indian, no other classical music:

- uses polyphonic harmony
- uses a rigid twelve note semi-tone scale
- uses composed, closed forms
- has such an elaborate and prescriptive notation
- has a machine as its chief instrument.

In China, Japan, India and the Arab world, the classic instruments of the elite music philosopher – *ch'in, koto, vina and 'ud* – are simple lutes or zithers. The classic studio instrument of the medieval European *musicus* and classical composer has been an elaborate keyboard mechanism – organ, harpsichord or piano. The same kind of dramatic difference between the instruments of music is to be found also in the musical materials. Apart from Africa, where loop polyphonies are known in some folk traditions, the great traditions have made music by the exploration of the acoustic fact of interval, in essentially single line or monodic music. Since the 11th century, within Europe, art music has been about combining two or more lines according to rules of composition that progress towards a resolution or close.

The postulate that music is the most immediate of the arts embodies the fact that through the tympanum membrane of the ear and the bone and cartilage articulation behind it, music literally strikes the senses. While it may be conceptually true that photons fall upon the optic nerve, I doubt anyone would claim to be able to feel the impact. By contrast a great church organ at full throttle can literally shake the building. The physicality of the art is part of the explanation of the fact that it has long been considered the most powerful of the arts. Plato was so nervous of its psychic power that he excluded certain of the musical modes from his ideal republic because he believed that they encouraged subversion and immorality.

Ancient Greek tradition held that the legendary hero musician Orpheus had once bewitched the god of the underworld. Similar tales come from all the great monodic traditions. A court musician to the Mogul emperor Akbar, it is said, came near to killing himself with his own music. Commanded by the emperor to perform the lamp light *raga*, he performed to such effect that he conjured up a real fire which threatened to consume him. Fortunately the considerate emperor called for a performance of the *raga* associated with rain and the flames were doused.

In such traditions the instrument itself could be accorded almost greater reverence than the performer. When, in 1603, the Italian Jesuit missionary, Matteo Ricci, finally gained access to the imperial palace at Peking, among his presents for the emperor was a clavichord. The emperor's musicians, who believed that the *ch'in* zither of their own tradition had been a gift of heaven, did obeisance to the little keyboard, supposing that it held the same aura for Ricci as the *ch'in* held for them.

It is to this world tradition that the music of ancient Greece and Rome belongs. Like Ricci's friends at the Chinese court, or like Indian musicians, who believed the *vina* to have been given by the goddess Saraswati, goddess of learning and the arts, Orpheus was believed to have received his instrument from the god Apollo. The bardic lyre to which Homer would chant his epic poems, had been the gift of heaven. With respect to his instrument, the art of the musician in a monodic tradition consists in exploration and exposition of the sound world that lives in the mystic body of wood, gourd or shell strung with gut. Compared with the keyboard and the festooned mechanisms of western wind instruments, these were organic members of the natural world. Modifications of their structure were, if not blasphemous, then pointless since the vocation of the musician was to enter and understand their musical world and not to 'improve' it.

Through the musician, moreover, human kind could approach the music of the spheres, the imagined musical structure of the world. The acoustic mysteries of interval and mathematics were linked to the secrets of the universe. The almost innumerable gradations of vibrational pitches between the unison and the octave were ordered into complex scales, Arab *maqqam,* Indian *ragas* and Greek *tropoi* ('tropes', the Greek 'modes' as they are called in modern theory), each held to embody and express its own microcosm of sentiment, mysticism or expression. Most systems recognized what we may call basic seven note scales (comparable with the white notes of the piano) coloured with semitone or smaller 'chromatic' (Greek, *k'roma*, 'colour') intervals. But this did not exclude the use of a cloud of other notes and intervals. Classical Indian theory for example divides the scale into 22 *shruti,* from which seven were selected and differently combined to build the *ragas*, each with their own expressive world with special significances as to time, colour, emotion, mood, sexuality and so forth.

But whereas the music of Orpheus and his lyre belonged in this musical ecology, the mechanistic music of keyboard and fixed scale of organ and harpsichord, of concert grand or rock keyboardist and folk accordionist do not. Theirs is another style of acoustic experience where the subtleties of fractional pitch modifications and intervals which can be appreciated in single line, i.e. monophonic, music, have been displaced by the increasingly rigid tunings of the twelve note semitone scale, adapted to a music of many lines sounding together, i.e. polyphony, and in which the human psyche found itself opening to another mode of life, the mode of machinery and progress.

In performance, too, the western art of composed music differs radically from the improvisatory forms typical of, for example, Indian classical music. The western concept of music is radically different. In the words of Stravinsky : 'The phenomenon of music is given to us to [establish] the coordination between man and time. To be put into practice, its indispensable and single requirement is construction. Construction once completed this order has been attained and there is nothing more to be said.'[1] A western composition is a closed form with designed climaxes and a set structure and conclusion. Things are different elsewhere. A classic exposition of an Indian *raga* may run for up to an hour or more if the performers and the listeners, who are an essential part of the event, are in sympathy. In a less sympathetic ambiance, the same musicians may expound the same *raga* in as little as twenty minutes.

Indian musicians have an apprenticeship of many years, learning not only the techniques of their instrument, but also an immense traditional vocabulary of actual musical phrases, to be used according to their own mood and genius in extemporary performances of the various ragas. Tradition is the watchword in non-European traditions as a whole – it is claimed that there are elements of Japanese court music which have remained in regular performance unchanged since the eleventh century. For at least as long, the European tradition by contrast has been devoted to innovation. The evolution of part singing and thence harmony, the development of a highly prescriptive notation and the adoption of mechanical instruments, above all the organ, reshaped not only music making in the west but also, as I shall argue, the very nature of human mentality in that tradition.

The first organ seems to have been the hydraulis invented in Alexandria in the 3rd century BC. It comprised four elements: ranks of pipes tuned in different pitches; some form of pump to blow the air to sound these pipes; a mechanism for controlling the flow of air to the pipes; and a method of maintaining a steady air pressure by blowing the air into a reserve chamber where the level was kept constant by hydraulic pressure. (The instrument's name comes from the Greek word for water and *aulos*, a reed pipe). As early as the first century AD, the mechanism to control the flow of air to the pipes appears to have been operated from a keyboard, possibly the invention of Hero of Alexandria.

Used in theatres, in state ceremonials, at weddings and above all in the Roman circus, where its music accompanied gladiatorial combats, this ancient organ was never an instrument for the highest musical expression. This remained the province of the voice and lyre, the instrument of the bard. However, even after the advent of the Christian emperors with Constantine in the early 4th century, the hydraulis organ continued to hold a central place in imperial ceremonial and in the hippodrome at Constantinople, where the rival chariot factions of blues and greens each maintained their own instrument and virtuoso performer – a kind of cheer leader keyboardist! Such were the antecedents of the instrument which in the Catholic west, though not in the Orthodox east, was to be admitted into the worship of the church. And it was the centuries of church culture which determined the shape and direction of western European civilization.

It is easy to forget that today's secular culture, like every other human culture we know of, was raised from a matrix of religious belief and practice. It was the church and churchmen who, in post Roman, barbarian ruled Europe were the carriers and the first guardians of literate culture. From the start, however, this western religious tradition was unusual. First, music had a central role. Secondly, that music involved the participation of the worshipers as well as the priests. As early as the second century of the Christian era, and possibly before, communal hymn singing was a focus of meetings among believers of all ages and both sexes. Because of the different pitches of the voices this was to lead to part (at first two – part) singing which would become known as organum. And if, over the centuries, this 'audience participation' became the province of a specialist minority group of choristers, the principle of group participation was maintained and the music itself, a choral music involving voices differing in pitch, evolved characteristics with special acoustic implications which in turn altered western musical sensibilities. The communal hymns of the primitive church, and the elaborate polyphonic music which evolved from them, were a development new in the history of world music and opened new patterns of thought in the minds of participants. Coupled with this was the arrival in the service of the church of a mechanical instrument capable of performing two or more parts moving simultaneously. Scholars dispute any possible connection between the term 'organum' to denote the style of part singing and the use of the word to denote the instrument. However, contemporaries had no such doubts. A 12th century French bishop, on hearing the organ for the first time, compared the sound to that of men and boys singing together.

The arrival of the organ in the west was due, in the first instance, to political and social rather than musical factors. The hippodrome instruments of Constantinople were not the only ones to be found in the great imperial city – indeed, probably the finest was the one kept in the throne room of the palace. There, together with various automata such as golden roaring lions at the emperor's feet, its function was to overawe visiting barbarian ambassadors, like those from Frankish Europe. On approaching the presence the intimidated envoy was required to prostrate himself in an obeisance reminiscent of the Chinese kow tow. On raising his head he was astonished to find that the enthroned emperor had been mysteriously raised several feet in the air. It is not recorded that the hydraulis player shared in the elevation (his day would come in the local Roxy or Odeon cinema of the 1930s), but his music, if not high art, was essential to the trappings of imperial prestige.

Sometime in the mid 8th century the king of the Franks received from the imperial court at Constantinople, in a gesture of prestige diplomacy, a hydraulis. The instrument appears to have broken down. At any rate nothing more is heard of it. Then, at Rome, on Christmas Day 800 the Frankish king, Charlemagne, was crowned emperor in the West. What was seen as a usurpation in Constantinople was regarded with pride by courtiers at the royal residence of Aachen. When about 820, Charlemagne's son and successor, Louis the Pious, acquired a new court hydraulis, built to commission by a Byzantine priest living in Venice, his pretensions were matched in the pageantry of the palace. Wrote one court eulogist : 'now the organ, the only reason why they of Constantinople thought themselves superior to you, oh Caesar, is at the court of Aachen.' It was, no doubt, a triumph in the world of diplomacy but had the instrument remained in the throne room it would also have remained, like the ones at Constantinople, a bizarre if impressive curiosity. The bards who entertained the nobility and emperor with the Lay of Roland and the sagas and epics which provided the focus of Germanic aristocratic culture had no interest in the contraption. But then theirs was the culture of the aristocratic elite. Only when it was taken up by the elite of the church, whose constituency was the whole Christian community, would its revolutionary potential be realized.

The move came in the mid 9th century. Just as the Aachen hydraulis had been built under the supervision of a priest, so the maintenance of the instrument according to the tuning ratios specified in the musical treatise of Boethius was consigned to church men – Europe's educated elite. Then about the year 850 Pope John VIII wrote asking that an instrument be sent to Rome. The papal court saw itself as the success – or to the Rome of the Caesars and the imperial music machine would have been quite at home. It was also seen as a useful teaching aid to choristers in the developing style of part singing known as organum.

By degrees, the organ was given a limited role in the liturgy. Organs were especially favoured in the monastic churches which followed the reforms introduced at Cluny from 911. The reformers believed that the worship of god should be conducted with solemnity and splendour, with buildings, vestments and also music of the highest quality. By the year 1000 organs were to be found all over Europe. Instruments were of increasing elaboration. It was said that an English example had no fewer than 26 bellows and two sets of keys and was audible a mile a away. The installation of any great church organ must have made a profound impression on the community. For months the church and its precincts were a construction site in which the pipes and apparatus of all kinds were open to curious passers-by. There were puritanically minded churchmen who objected to such machines as a distraction from the service of god. In the 12th century, the Yorkshireman St Ailred of Rievaux fulminated against the arcane complexities of modern music, designed in his view to show off the talents of the musicians rather than exalt the glories of God, and against the organ and its operator whose activities transformed the church into a place of entertainment and distracted the gawping congregation from the rituals being conducted by the priest. But most of the clergy accepted the innovation – one French bishop actually had the organ of a rival removed from a neighbouring church to have it installed in his own cathedral.

Thus, uniquely in world civilization, the European tradition combined the two most powerful impulses of early culture, music and religion, to the service of the mechanical principle. Beside the organ, 11th century Europe boasted another large music machine, the 'organistrum', in which the sound was produced by drone and melody strings vibrated by a rosined wooden wheel operated with a crank handle. Like the organ it sounded two or three notes simultaneously, like the organ it could be used as an aid in the choir school and, like the organ, it was for a time adopted in parts of the liturgy. Once again, the priestly caste admitted the machine into the sanctuary.

The organistrum had a comparatively short life as a church instrument. By contrast, its derivative, the one man hurdygurdy, was to carry the concept of machine music into every stratum of European society. In the hands of jongleurs and travelling minstrels it could be heard in the halls of the aristocracy or at peasant weddings, while at the performances of miracle and mystery plays by town guilds it provided the music of the orchestra beneath the pageant cart stage in busy city streets. European folk music, like European art music, was the only one in the world to accept a mechanical instrument. The hurdy gurdy continued a part of the living folk tradition e.g. in the Auvergne, into the early 19th century when it was generally displaced by the improved mechanism of the accordion. It is worth noting that for Lynn White Jnr. the hurdy gurdy offered persuasive proof that the crank appeared in Europe as the result of independent invention rather than transmission from China. 'The crank,' he wrote, 'is a mechanical element which can spread only as part of a larger device, and it appears first in Europe on the rotary grindstone and then the hurdygurdy, neither of which was used in China.'[2] Whether or not the argumentation is sound, it is entirely characteristic of western culture that a musical instrument can be adduced as a piece of primary evidence in a thesis on the history of technology.

By 1100 the practice of singing in two parts moving together was becoming more elaborate and soon a third part would be added. Such developments created problems for choirs and choir masters looking for well coordinated ensemble singing. Up to this time musical notation, such as it was, consisted of mnemonic marks rather than precise indications of varying musical pitches. In the middle of the 11th century the Italian choir master Guido d'Arezzo introduced a four-line grid or stave on which notes were represented by diamond shaped marks. Each move up the grid, from a line to the space above it, represented a fixed interval, usually a single tone. Guido's system, the original musical stave, was rapidly adopted. With its aid, choirmasters could be sure that properly trained choristers would sing their parts at the correct pitches and, equally important, even young singers could be taught new music more quickly and surely than before.

Guido's system anticipated the European fascination with labour saving devices, and this in the one area which in other cultures was almost intended to be laborious. Where the Asian musician expects to spend years memorizing the vocabulary of his art, the choir at Arezzo and soon the choirs of all Europe's cathedrals, freed from that particular chore, were liberated for the exploration of new music.

While Guido had provided a simple and effective codification of pitch, the other main variable in the music of his day, rhythm, presented singers with ever more difficulties as new music became increasingly elaborate. By 1300, according to the contemporary English theorist Walter Odington, there were as many systems of notating rhythm as there were theorists. It was the French mathematician and musician Philippe de Vitry who, in a series of lectures delivered in Paris in 1316, outlined a system which did for rhythmic notation what Guido had done for pitch. It became known as the *ars nova*, a term which had none of the mystique of 'art' as understood by the renaissance but meant simply the new technique or craft. The early evolution of effective systems of musical notation, by making possible more reliable records of past practice, liberated the musician from much memorizing and so encouraged creative innovation.

Whereas in other cultures, classical musicians were the guardians of tradition, in the west they have always been in the vanguard of innovation; Guido's new notation was followed at Notre Dame in the 12th century with new organum compositions by Leoninus and Perotinus. The *ars nova* of Vitry was followed in the 15th century by what continentals regarded as the new style of the Englishmen Dunstable and Leonel Power, the *contenance angloise* and thereafter every century had its proudly proclaimed new styles in music. In the 1780s Haydn published quartets which he described as written on new principles. The New Music of Wagner provoked his critics to near apoplexy and in 1962 America's Fromm Music Foundation established

'Perspectives of New Music.' What the art critic Theodore Rosen called the 'Tradition of the New' is virtually as old as the new European culture which developed after the fall of the Roman empire in the West and nowhere stronger or longer established than in its art music, the art which is the most immediate to the electrical impulses of the brain and which in other cultures is most closely associated with tradition and continuity.

But the greatest of all western innovation in music is, that majestical web of structured sound known as diatonic harmony. As has been indicated, it inhabits an audio world different in kind to the monodic and intervallic concerns of the other great traditions of music – in fact it created that world, a world in which form is closed but dynamic and above all progressive in the sense of moving forward to a sensed objective or resolution. In English, this art of forward movement is betrayed in its very vocabulary. One speaks of the 'leading note' which impels towards rest and resolution ; harmonies change by 'modulation' and chord 'progressions'. This progressive art, evolved by Europe's musicians and to be heard by all who attended church in the centuries when church attendance was conventional, infected and reinforced western thought patterns. Thus, in western Europe, cradle of modern world culture, the practitioners of the most immediate of the arts established the habit of innovation and progress thinking as part of the cultural mindset and were sanctioned by the most powerful of all traditional values – religion.

Until well into the 1500s musicians from northern Europe, Paris, the Low Countries and briefly England, dominated the art. Their innovative and progressive music was the antithesis of ideas developed by the Italian humanists in the 15th century. The humanist ideal was a return to all things ancient:

- Return to ancient classical architecture in place of the dynamic engineering of the Gothic centuries
- Return to Cicero's Latin in place of the everyday working language used by medieval scholars and churchmen
- Return, of course, to the monophonic music of ancient Greece with its modes.

By tradition this had been learnt in the works of the 6th century Roman philosopher Boethius. His *de musica* provided the basic text book for musical theory well into the middle ages. But as early as the 13th century this treatise in the monodic tradition of the classical past had about as much relevance to practical music making as an 18th century book on harmony would have to a composer of twelve note music in the 20th. Theorists, of course, continued to advocate its precepts, but for practising musicians it was an academic irrelevance.

But even the theorists had their difficulties with Boethius. Some 14th century treatises make one wonder how much of his work was understood or even studied at first hand in that century of swiftly changing musical practice. Things could only get worse and the complexities of the situation are encapsulated in the work of the 15th century Flemish musical encyclopaedist Tinctoris, a master exponent of theory but equally a devotee of practical contemporary music. In his *Liber ... tonorum* of 1477, he doest his best to relate the multifarious polyphony of his own time to the venerable tradition of the modes, but he also 'states clearly that mode is attributable to a single melodic line only'.[3]

The careful study of Boethius returned with the humanists. For they were rarely pracitising musicians and, unlike Tinctoris, had only contempt for modern music. What they wanted was the revival of a dead tradition. Unfortunately for them, there was no sure way of finding out exactly how ancient Greek and Roman music had sounded. New-agers before their time, these antiquarian musicologists evolved a complex system combining philosophy, magic and the use of talismans with the practice of what they called orphic songs to draw down the influence of the planets. They wished to revive a monodic tradition abandoned in western art music half a millenium before. It was a lost cause. Such practising musicians as did admire their theorizings, like Vicenzo Galilei, father of the mathematician, astronomer and physicist, wrote their finest works in the prevailing polyphonic idiom of their contemporaries. The very instrument the devotee theorists invented to accompany their supposed Orphic monodies, betrayed their ideal. In its name, the *lyra moderna*, it consciously evoked the harplike instrument of Orpheus. But where the Greek lyre had been plucked, the 'modern lyre', something like a largeish five stringed violin, was bowed and its virtuosi astounded their hearers 'because [they] were able to play four parts at once with lightness and musically.' So, even the most dedicated advocates of reviving the forgotten monodies of the ancients were seduced by the burgeoning polyphonic idiom of their own day. Their struggle to reverse the current of musical evolution established by their 'medieval' predecessors was doomed – the course of progressive innovation was unstoppable.

Finally, in this skeleton survey of the musical mindset which established in Europe the technoculture of the modern world, I shall briefly comment on the remarkable saga of invention which is the history of the keyboard instruments. At every stage that development was in response to a specific problem of music making, analyzed and dealt with in classic problem solving terms. The sequence, which I have described in detail elsewhere, is best compared to a project in research and development.

The earliest organ keyboards, dating probably from the ninth century, had seven notes for each octave, sounding pipes of the major scale like the white keys of a piano. However, new notes were added to this simple 'diatonic' scale of tones and two semitones. By the year 1300 the familiar scale of twelve semitones was already developed and instrument builders were faced with the problem of designing keys such that twelve could be controlled in the span of a hand. One solution found in small organs was to use typewriter-like stud keys mounted in two rows. Other makers used split keys similar to the ones on the piano and this became the preferred system. The next challenge came when some ingenious mind turned its attention to fitting a keyboard mechanism to strings rather than pipes. We first hear of an instrument of strings played like an organ in the mid 14th century. It was called the *eschiquier anglais*. No examples survive, but it was probably a form of *clavichord*

(Latin claves, a key), developed by the end of the century. In this instrument a series of pivoted lever keys with metal tongues or 'tangents' let in to the end sounded the strings running parallel to the keyboard when the finger key was depressed so forcing up the further end with the tangent which thus struck and sounded the string. It was an ingenious adaptation of the keyboard familiar on the organ but fitted with tongues similar to those found on the hurdy-gurdy. The mechanism was simple and the musical effect delicate however it was neither loud nor brilliant, effects found on the plucked stringed instruments. By 1404 some mechanical genius had solved the problem of how to make the keyboard operate a plucking mechanism and the first member of the harpsichord family was born. During the 16th century keyboards were designed. They enabled the player to control yet more strings tuned to microtones permitted the instrument builder to mount the strings vertically against a wall so as to save space, to operate strings sounded by revolving bow wheels (a kind of keyboarded viol, Leonardo's idea this), among others. Then, in the 1710s, the Italian Christophori produced his *clavicembalo pian e forte*, a keyboard mechanism which projected hammers against the strings. With this the musician was severed from the direct contact with his instrument which for the bard had meant an almost mystic communion with the world of the gods and the music of the spheres. For at the moment, when the sound is produced from the string, the moment when the musician's touch vibrated the fabric of the godgiven instrument, the string is in fact sounded by a hammerhead projectile in free flight.

I have mentioned just a few of the plethora of keyboard innovations and adaptations which throughout the history of western music have provided classic expression and exercise for the inventive and innovative genius of the western mind. In the 19th century this millenium of R & D shifted gear. In 1837 another Italian, Guiseppe Ravizza, adapted the musical keyboard to mechanical writing, calling his invention the *cembalo scrivano*. In 1870 came the Remington company's patent for the first commercially successful typewriter. In our own century, with the keyboard's application to computing, the studio tool of the medieval *musicus* or scholar musician has entered the service of the data bank. The keyboard is the most easily observable, though by no means the most profound, aspect of the way in which the western musical tradition has shifted the human psyche to a mindset in which progressive innovative technology has become the dynamic of human culture.

Notes

This is a revised version of an article published in *ICON. Journal of the International Committee for the History of Technology* 1997, 3:167-180.

1 I. Stravinsky, *Chronicle of My Life* (London, 1936), 91.

2 L. White Jr., *Medieval Technology and Social Change* (London, 1965), 97.

3 G. Reese, *Music in the Renaissance* (London, 1954), 141.

Hugh Davies

ELECTRONIC INSTRUMENTS: CLASSIFICATIONS AND MECHANISMS[1]

In 1914 Erich von Hornbostel and Curt Sachs published the first systematic classification system for musical instruments. Although a handful of electromechanical instruments existed at that date and would certainly have been familiar to them from brief descriptions published in journals such as the *Zeitschrift für Instrumentenbau*, these forerunners of the major instrumental innovations in the twentieth century were not included in their classification system. Indeed any attempt to do so, based on such a small selection of the total possibilities that were subsequently explored, would have been a failure. The electronic oscillator, which was developed very slightly later and has become the most common method of sound generation in this area, is indeed the only new principle of sound production that has been discovered since prehistoric times. In spite of the astonishing figure of one billion mouth organs manufactured up to 1987 by Hohner alone, it is probable that more electronic instruments than acoustic ones are being manufactured today, and before long it is likely that more electronic instruments will have been built than all the acoustic instruments made throughout human history, particularly if one includes such primarily non-musical devices as pocket calculators, digital watches, home computers and electronic games machines (all of which can produce a selection of electronically generated or sampled melodies or other sounds), as well as special types of greetings cards and books and toys for young children. The recent growth of musical applications for the microcomputer can only increase in the next few years; certainly it is already well supported by software programs, new types of 'black box' and a variety of specialist books and magazines.

Whatever each person feels about the sound qualities available on, and the music created with, today's digital synthesizers and microcomputers, there can be no doubt that they are here to stay: the advanced electrical technology on which they are based is the only area in which prices are constantly reduced, because the ubiquitous microchip which forms their central element has become so cheap to manufacture (despite the laboratory-like dust-free conditions required for this), made as it is from such common materials as sand (silicon). The production of ever-more sophisticated devices is eagerly supported by the large rock music industry, which ultimately owes its existence and especially its growth in the last 30 years to one thing: the parallel increase in the spending power of children's pocket money!

The spread of the synthesizer during the period of 35 years since the first ones were commercially manufactured has been astonishing, rivalling the popularity of the electric guitar, even though, over the last 15 years, it has been surpassed by that of the microcomputer. Synthesizers and samplers are used in all kinds of music, especially in the more popular and commercial areas, but they are now beginning to be included in works by younger composers (mostly with two digital synthesizers) for symphony orchestra, which, because of its repertoire of specially composed music from a period of more than two hundred years and the salaried status of its members, is the last bastion to be surmounted. No such resistance to introducing synthesizers exists in rock groups, dance bands and other ensembles where the repertoire is specifically arranged if not also composed for the particular combination of instruments available at the time. Furthermore, the spread of the synthesizer has not been limited to parts of the world that are fully Westernised: since the early 1980s it has become common in Indian film scores and West African popular music. In the same period local commercial production started in Eastern European countries, especially East Germany, Poland and the Soviet Union, where small electronic organs and electric pianos had been manufactured since the 1950s on a similar basis to that in Western Europe, and imported synthesizers have been played by rock, jazz and other groups since the mid-1970s.

CLASSIFICATION: ELECTROPHONE

Music designed to be heard over loudspeakers (including not only all music for electronic instruments but also taped, live electronic and computer music) is commonly known as 'electronic music' or 'electroacoustic music'. This choice of terminology is based on the selection of one or other of the two extremes within the area to describe the whole. The word 'electronic' is used here to define all the instruments that for classification purposes are known as electrophones. It includes those instruments, of which the best-known is the electric guitar, in which electrical amplification is an essential part of the process of sound production, and must thus, in terms of physics, be considered as electrophones. Such instruments are normally called 'amplified' or 'electric' in everyday conversation by both musicians and electrical engineers. For classification purposes, however, it will be seen that a more precise definition and description is required of the way in which the sound is created, and for which the term 'electroacoustic' is used.

There are three basic forms of electrophone, each of which can be further subdivided. The term 'electrophone' appears to have been first introduced by Sachs himself in 1940, in *The History of Musical Instruments*, where he divided electronic instruments into 'electromechanical' and 'radioelectric'; the present three forms were first proposed by Canon Francis Galpin in 1937, in *A Textbook of European Musical Instruments*, and modified by him a few months later in a lecture 'The Music of Electricity' (given at the age of 80), as 'autophonic', 'mechanical' and 'acoustical'. These, renamed in the light of subsequent knowledge and understanding, are:

1) electronic instruments, in which the sounds are generated by one or more analogue or digital electronic oscillators or different methods of digital synthesis

that achieve equivalent results. Subdivisions can be made in terms of the type of analogue oscillator circuit employed, the oscillator's waveform (usually closely linked), the number of oscillators required to produce each sounding note and whether the pitches are generated at the same frequency or derived from a higher (occasionally lower) one by means of difference tones (beat-frequency oscillator) or by frequency division (or occasionally multiplication);

2) electromechanical instruments, in which waveform shapes are usually inscribed on one or more continuously moving strips (film, magnetic tape) or rotating 'tone-wheels' (discs or cylinders). These are subdivided according to the method in which these shapes create a varying voltage at one or more frequencies: electromagnetic, electrostatic or photoelectric. Such electromagnetic systems are equivalent to one or more electronic oscillator circuits;

3) electroacoustic instruments, in which the vibrations of one or more objects such as strings, reeds, plates or rods (which may have their acoustic audibility reduced or eliminated by being enclosed or by having no resonator, as with the solid body of the electric guitar or the lack of a soundboard in most electric pianos) create a varying voltage at one or more frequencies by means of special built-in transducers (which are usually but not invariably a type of microphone) based on one of the following principles – electromagnetic, electrostatic, photoelectric or piezoelectric (pressure-sensitive). Such electroacoustic systems are equivalent to one or more electronic oscillator circuits.

The principal distinguishing features of the three types of oscillator are shown in the following table:

	Moving parts	Oscillator type
electronic	No	Active
electromechanic	Yes	Active
electroacoustic	Yes	Passive*

* A passive oscillator is one in which no electrical power supply is necessary, since, unlike the other two categories, the oscillator does not permanently operate whenever the instrument is switched on, and when it is activated it generates the necessary voltage to affect the next stage in the chain of electrical equipment, the amplifier. In a few cases, especially with capacitative and electrostatic transducers, an electrical power supply is an essential adjunct to the human energising element.

The von Hornbostel-Sachs classification system places conflicting emphases on the method in which a vibrating object is activated. With idiophones and membranophones it is the most important subdivision of each category, while with cordophones and aerophones it is treated as the smallest possible distinction, as with the use of the bow (indicated by the final digit in the Dewey-based numerical classification) that separates a violin from a guitar, or whether a nail violin is bowed, plucked, struck or even blown on – in other words, the performer's side is of minor importance compared with that of the sound generation. In everyday use an electroacoustic oscillator is normally seen from the musician's viewpoint and described as 'amplified', but for classification purposes it must be seen from the viewpoint of the chain of equipment on the electrical side (amplifier, loudspeaker), in which the vibrating object is an oscillating frequency controller that is the all-important first stage.

When studying the application of electricity to musical instruments it is important to determine where it is utilised in the often complex chain of three stages that are necessary to produce a sound, just as in conventional instruments: activation, (mechanical) 'action' and sound generation. In the simplest case, manual activation of the sounding object, such as a guitar string, includes the action; in the electric guitar, electricity is used only in the sound generation stage. Keyboard instruments, especially the piano and pipe organ, are somewhat more complex, since the action stage is mechanical, or, in many organs of the last 100 years, electrical. With electric action, electromechanical switches transmit changes from the activation stage to the sound generation stage, as first used for bell tremoli in the electrostatic Clavecin électrique (1759), inspired by a simple detector system for the presence of an electrical charge used by early electrical experimenters. Electric action might appear to be essential for all electromechanical and electronic instruments, but there are a few exceptions: 'space control', in which the performer does not touch the instrument, as with the Theremin's hand capacitance (developed from the inventor's previous intruder detector) or hand and body movements between a light source and a photocell (as in Qubais Reed Ghazala's *Photon clarinet*, several installations by Christopher Janney, beginning with *Soundstair*, and Jacques Serrano's computer-controlled *Mur interactif spatio-temporel*) or in an infra-red beam (Don Buchla's *Lightning* and Interactive Light's *Dimension Beam*, licensed by Roland and incorporated in several recent synthesizers as the *D Beam*), low-frequency radio waves (with special drumsticks in the *Radio Drum* of Max Mathews and Robert Boie and batons in Mathew's *Radio Baton*), the distance-sensing airdrums, and oscillator frequencies modified by body resistance (Michel Waisvisz's 'crackle' system and 'The Hands'). On the other hand it can also include electromagnetic coils and/or rotating tone-wheels that cause objects such as strings to vibrate without direct contact, as in Richard Eisenmann's *Elektrophonisches Klavier* (c.1885-1913) and the Choralcello (c.1888-1908). Apart from pitch controls there may be additional independent activation stages with separate actions, involving registration, loudness or vibrato controls, as in the electrically motorised vibrato of the vibraphone, and loudspeaker vibrato in which either the loudspeaker rotates (as in the Leslie) or a diffusion 'paddle' is rotated in front of it (the Everett Orgatron).

The main activation stage may also be automated by using a mechanical (as with earlier mechanical instruments), electromechanical or electronic control system that replaces the human performer. It would be possible – if somewhat absurd – to place a *Vorsetzer* piano player, controlled by a punched paper tape, in front of a digital synthesiser; but there are modern forms of the player piano, such as Marantz' digital Pianocorder (1978) and the computer-controlled Bösendorfer grand piano and Yamaha Disklavier from ten years later. Electronic forms of 'memory' storage for passages of music (as well as for subsidiary activation stages) include a variety of systems: electromagnetically in the form of analogue or digital magnetic tapes or magnetic discs (as with computers and digital synthesizers), photoelectrically with optical discs (as in the Mattel Optigan) or – using masks – on cards, sprocketed film or rotating tone-wheel discs (similar to, but separate in function from those used to generate sound), and electronically in the form of plug-in digital storage devices like RAM and ROM cartridges (as in the Yamaha DX7) and compact discs (CDs; including CD-ROMs and the newly available re-recordable CDs).

A final consideration in this area is whether an instrument is monophonic, partially polyphonic or fully polyphonic. Traditional polyphonic instruments include the harp, piano, pipe organ, piano accordion and keyed percussion, which make all pitches of the tempered semitone scale within a specific pitch range simultaneously available to the performer. Partially polyphonic instruments are those in which only a limited number of the pitches within the overall range are available simultaneously, as with the guitar and violin families. Such a distinction is rarely made with acoustic instruments, but it is valuable for defining a few early electronic organs and many polyphonic synthesizers, which contradict the fully polyphonic nature of nearly all previous keyboard instruments (apart from the monophonic keyboard of the hurdy-gurdy and all monophonic synthesizers) by providing a maximum number of keys (such as eight or 16) that can function simultaneously. A selection system ensures that this number is not exceeded, using as a criterion for elimination the fact that a key was the first or last, or the highest or lowest, to be depressed. Other related features, always with a historical precedent, include the 'split' keyboard (as on some harmoniums) and the system of 'shared oscillators' found in some earlier small electronic organs, in which two or – more rarely – three adjacent keys controlled a single oscillator and could thus not be sounded simultaneously (as with early 'fretted' clavichords).

Electroacoustic Instruments

Under the classification heading it was convenient to begin with electronic instruments, in order to show how the other two categories, the electromechanical and electroacoustic, are intermediate stages which also involve equivalent forms of oscillator that incorporate a mechanical element. In order to discuss the different possibilities in each category the reverse order has been chosen, since it reflects the actual chronological development more closely.

The passive electroacoustic oscillator normally consists only of a vibrating object that is positioned close to one or more electrical coils, in between a light source and a photoelectric cell or in direct contact with a piezoelectric crystal. It is equivalent to connecting the 'output' of a hand-cranked low voltage generator (as used for field telephones) to an amplifier and loudspeaker for use as a low frequency oscillator, which may sound improbable but is perfectly feasible. The type of electroacoustic oscillator that uses the electrostatic principle, although incorporating a vibrating object that permanently carries a voltage, is also passive, since there are no permanently moving parts; the object (such as a string, reed, plate or rod) is no more than a 'straight length of wire' within an electrical circuit (and thus carrying a voltage) until it is set into vibration by means of the appropriate part of the instrument's activation stage. The electrostatic oscillator is then set in motion, affecting one or more coils in the vicinity of the object, as with an electromagnetic oscillator in the same category.

Most electroacoustic instruments closely resemble their acoustic ancestors, such as pianos, harmoniums or reed organs, carillons, guitars and bowed string instruments. Since we require that there should not be a basic aural difference between what is heard over a loudspeaker from an instrument with built-in microphones and the sound that results from placing a standard 'air' microphone in front of the equivalent acoustic instrument, it is initially difficult to convey the subtle distinction that the amplified result is not produced in an identical manner to the purely acoustic sound, but only in one that is parallel to it. There are, however, certain types of transducers that function in a significantly different way, notably the electrostatic one (rarely if ever used today), in which, although there may not be any easily visible difference between it and an electromagnetic transducer (or in any associated circuitry), the electrical situation is completely different. Furthermore, contemporary digital technology permits a transducer or sensor to provide information for a microprocessor (as on the hi-tech SynthAxe and Stepp digital guitar-like synthesizer controllers) which can be instructed, for example, to adjust the frequencies produced on any one or more strings so that they no longer correspond to those that would be heard if the strings could be amplified in the normal way. An electroacoustic instrument can be recognised as such when it is clear that it is the designer's intention for it to be heard with amplification, by incorporating one or more microphones or other transducers in it, normally in precisely determined positions, and usually by significantly reducing the loudness of the sounds that can be heard acoustically; to ignore this intention is equivalent to turning a grand piano upside down and claiming that it is a frame drum with sympathetic strings!

Photoelectric instruments in this category, in which the movements of a vibrating sound source mask a beam of light that is received by a photocell, have been very uncommon; this may change as a result of the recent growth in specialised forms of light, such as laser and infra-red beams and fibre optics with appropriate sensors. Physical contact between the sound source and the transducer is also not required in electrostatically and electromagnetically amplified instruments. The electrostatic principle, first introduced in the 1920s and featured in several musical instruments in the next few years, was mentioned above. The electrical side of the transducer is usually a rectangular bar or plate which functions as one electrode in a variable condenser, whose other electrode is the vibrating sound source, usually separated from it by a distance of up to 1 cm. Similar in appearance and in its proximity to the vibrating source is the electromagnetic transducer or 'pickup', best known in the electric guitar, and also used in the Neo-Bechstein grand piano. Thin wire is coiled around a magnet, whose field is affected by the movements of any object close to it that contains sufficient iron, thus creating a varying electrical voltage. The earliest use of an electromagnetic pickup for a musical instrument appears to have been in 1874. The final type of transducer, based on the piezoelectric ceramic crystal (a principle that was discovered by Pierre and Jacques Curie in 1880, but not used successfully for a microphone until 1931), is the only one that requires direct physical contact with the vibrating sound source, since it is based on the ability of certain ceramic crystals to generate a voltage when stress is applied to them; a similar application of piezoelectric crystals is in ceramic record cartridges.

The range of electroacoustic instruments is extremely wide and can only be surveyed here briefly (more details can be found in the author's article 'Electronic instruments' in *The New Grove Dictionary of Musical Instruments*, 1984). Keyboard instruments include electric pianos (with strings, struck reeds, plucked reeds or struck rods, amplified by one of the three main types of transducer), harpsichords and a clavichord (the Clavinet). A few 'sustaining pianos' were constructed between the 1880s and the 1930s in which, when a key was depressed, the string (or other resonant object) was activated by an electromagnet instead of being struck by a a hammer, but most of them were not amplified; a recent example of the principle is the hand-held E-Bow, which activates individual strings on an electric guitar.

Experiments towards developing an electric guitar first took place in the early 1920s (electrostatic pickups were added to a guitar by the designer Lloyd Loar in 1923), primarily motivated by the need for an instrument capable of greater loudness in the jazz and popular music bands of the time, a role that the entirely acoustic dobro, with its inset metal resonator discs, pioneered from the late 1920s until World War II. During the 1930s Loar, Rickenbacker and others built electric guitars with electromagnetic pickups, while in 1941 the guitarist Les Paul produced the first solid-bodied instrument, known as 'The Log' because of its detachable non-functional side pieces. Commercial manufacture of the electric guitar on a large scale did not begin until the early 1950s, and it was not until thirty years later that the first fundamental redesign of the instrument was initiated, in the influential 'headless' model of the Steinberger company (subsequently also licenced to other companies, including Hohner). More radical modifications of the instrument include 'The Stick' (Emmett Chapman, 1973), in which a percussive playing technique that requires only the left hand is extended to both hands independently (similar to Stanley Jordan's style on the normal electric guitar), and instruments built by Glenn Branca, Fred Frith and Hans Reichel. Other electric plucked string instruments include Hawaiian and pedal steel guitars, banjos, mandolins, harps, autoharps and sitars, while other electric struck strings have included two early 'electronic timpani'.

Electric bowed string instruments, also often with solid or reduced bodies and amplified electromagnetically or piezoelectrically, have been constructed since the late 1920s, frequently by the same builders or companies that were developing electric guitars. In recent years the limited introduction of, in particular, violins and cellos into rock music and jazz has created a renewed market both for fully electric instruments and for effective attachable microphones (normally piezoelectric). Some performers in areas of music such as free improvisation have constructed or commissioned electric cellos and double basses, while each string in Max Mathews' quadraphonic electric violin is vibrationally isolated from its neighbours and separately amplified. By 1987 several commercial electric guitars and electric violins featured MIDI (Musical Instrument Digital Interface, 1983), a system that enables devices equipped with it to be linked so that one instrument can fully control a very different one.

Electroacoustic instruments that incorporate reeds (plucked, struck, blown or sucked) include not only some electric pianos but also reed organs (such as the early Orgatron) and the unusual Guitaret, in which the performer's fingers pluck zanza-like tongues which cause the reeds to vibrate. Electric keyed percussion such as glockenspiels and vibraphones have been manufactured, without any great popularity, while electric carillons, based on amplified bells, tubes, reeds, plates, bars, rods or springs, have been built since the 1930s, and manufactured with considerable success since the 1940s; for example, by the mid-1960s in the USA a single company had installed over 5000 electric models, while there were no more than 80 acoustic instruments in the whole country. Similar transducers are used as sensors in those recent electronic percussion instruments that are controlled by playing special drums, drum-pads or xylophone-like bars. Mechanical electroacoustic instruments include ones designed for sound-signalling (as with bells and sirens), such as the amplified music-box mechanisms (plucked steel reeds) installed in ice-cream vans and the amplified chimes (struck steel rods) that introduce a spoken announcement in some railway stations and airports. Microphones have been built into a considerable range of instruments invented by composers and performers, only some of which sound or look like familiar instruments. Finally, several companies have developed specialised contact microphones for amplifying the piano, or for each of the most commonly played acoustic instruments, and some acoustic pianos, in addition to electric ones, now include a MIDI interface.

Electromechanical Instruments

Whereas electroacoustic instruments contain resonant objects whose vibrations are not only converted via a transducer into audible sound but can also be heard acoustically (like the very quiet direct sounds from the grooves of a long-playing record), this is not the case with electromechanical instruments. In most of these the vibrating element is caused by a rotating cylinder or disc called a tone-wheel, which contains one or more waveshapes inscribed on its rim or in concentric bands on one side that affect an electromagnetic, electrostatic or photoelectric component in an electromechanical oscillator. Unlike the electroacoustic category, the photoelectric principle is of equal importance here, while the pressure-responsive piezoelectric one seems not to have been used. In the last few years cheaper and more efficient digital circuits have completely ousted tone-wheels (an exception is Jacques Dudon's *Synthétiseur photosonique*) and similar mechanisms, although homage is still paid in some recent electronic organs to the best known early tone-wheel instrument, the Hammond organ, by incorporating accurately synthesized imitations or 'sampled' (digitised) recordings of its actual sounds. This trend has grown substantially in the 1990s in the form of commercial sampled recordings on CD and CD-ROM of every imaginable Western instrument, and many non-Western ones. The electromechanical oscillator of the RCA Electronic Music Synthesizer (1950s) unusually incorporates a tuning fork continuously excited by electromagnetic coils; this might appear to be classifiable as electroacoustic, but in fact it is similar to a tone-wheel oscillator. The tuning-fork functions as an extremely stable frequency control like the crystals in very high frequency oscillators, such as those in radio transmitters.

In most tone-wheel instruments there is either a set of twelve concentrically banded cylinders or discs, one for each semitone within the octave and all its octave transpositions, or one disc for each note on the keyboard. The earliest such instrument, the enormous Telharmonium, had cylinders about 46 cm in diameter, while only 30 years later the discs (one for each note) in the Hammond organ had a diameter of less than 5 cm. Such discs are usually mounted on one or more shafts that are continuously rotated by means of a synchronous motor, and are made of metal (electromagnetic), plastic (such as Bakelite; electrostatic) or glass (photoelectric). In the first two types the waveform consists of a continuous regularly repeating waveshape which is either the outline of the tone-wheel's edge or marked on one of its side faces. A cylinder that is the equivalent of several such discs contains several such profiles, like the concentric waveform rings inscribed on the side of a disc; these produce either different frequencies (different numbers of complete waveform cycles, as with the Telharmonium, normally 7-8 separate bands of a single waveform, each for a different frequency), or different waveshapes for a single frequency – or, occasionally, both. Larger glass or transparent plastic discs (typically around 30-40 cm in diameter) containing concentric rings of shaped 'teeth' or other forms instead of single waveform outlines, are found in many photoelectric instruments, while a few use shaped holes cut into rotating opaque discs (as on a siren) or short lengths or loops of sprocketed film.

With photoelectric instruments the waveshapes function as masks that, based on the principle originally introduced in the late 1920s for the optical film soundtrack, modify the way in which a beam of light is received by a photoelectric cell, while the varying profile of the rim of a rotating electromagnetic disc or cylinder similarly affects the field of one or more electromagnets placed nearby. In the case of electrostatic tone-wheels (as in the Dereux organ and the Compton Electrone) a pair of opposed discs is required: the moving rotor disc contains electrodes that are about 1 mm distant from the protruding shapes of the waveforms engraved on the fixed stator disc. In some instruments the waveforms are sinusoidal, with appropriate overtones 'borrowed' from other discs, while in others more than one waveform for each note is provided on one or more discs; some of these more complex waveforms were derived from photographic images of well-known pipe organ registrations (especially in Edwin Welte's photoelectric *Lichtton-Orgel*), a forerunner of today's digital 'sampling' technique. This latter approach can also be seen in a variety of instruments based on prerecorded sounds played back by other kinds of apparatus, such as magnetized steel discs, steel wire and gramophone records (variously between 1920 and 1950), an approach which was only fully successful in the Mellotron (designed in the early 1960s), using the magnetised plastic tape and more stable tape transport systems developed for the tape recorder during and after World War II.

ELECTRONIC INSTRUMENTS

In electronic instruments the oscillator is fully electronic and contains no moving parts; it is based on one of the four stages in the evolution of electrical circuits that are now used to define the different generations of computers: valves, semiconductors, integrated circuits (ICs) and very large scale integration (VLSI). The waveforms range from pure sinewaves to the randomness of noise generators (for percussive sounds). The fluctuating electrical situation in such oscillators functions in a manner that is comparable to the regularly varied outline of the inscribed waveform shapes in electromechanical instruments or of the vibration patterns of the 'acoustic' sound sources in electroacoustic instruments; the circuits normally alternate between two states, one of which is an exact reversal or negative form of the other. Most of the earliest electronic instruments, which were almost all monophonic, were based on a sinewave produced by a beat-frequency oscillator; this actually consists of two (inaudible) very high frequency oscillators, one fixed and the other variable, with only the difference between their two frequencies falling into the audible range. Later on, especially in polyphonic instruments, oscillators generated their frequencies either 'at pitch' (for the larger instruments, one per note) or else, as in the Philicorda, a set of twelve high frequency oscillators provided the twelve semitones of the highest octave available on the instrument (or often, for practical reasons, one octave higher) which were frequency-divided to create all the notes in the lower octaves; in the 1970s the latter system was largely replaced by a single (inaudible) very high frequency oscillator which is digitally divided to create the highest twelve semitones and then further divided as before, and in the 1980s by a set of digital oscillators; in recent years sampled sounds have become predominant.

Monophonic instruments. This category includes many early electronic instruments, all the so-called piano attachments (such as the Solovox, Clavioline, Ondioline and Univox) and most early synthesizers. As with acoustic instruments, the range of methods of performing monophonic instruments is considerably larger than that for polyphonic instruments, and the facilities for controlling the timbre and articulation of each note are normally more substantial. Apart from those instruments that are based on conventional keyboards (often with a reduced compass, such as only three octaves), the other control methods involve fingerboards (especially popular with some earlier instruments in the Soviet Union), rotatable dials (primarily for two early French instruments) and space-control (as in the Theremin and similar instruments from the 1920s and early 1930s). 'Fingerboard' refers to a narrow strip or ribbon (like a string above the fingerboard on a bowed string instrument), on which the performer places a finger to select different pitches, the position of the finger on the string or ribbon determining the pitch; in this case, the frequency of an electronic oscillator (or, in some synthesizers, controlling a sound-modifying device such as a filter). The ribbon is often placed transversally in front of the performer and fingered like a keyboard, with which it is indeed sometimes combined – as with the Ondes Martenot – primarily to provide glissando and portamento effects that are otherwise unavailable to the performer. Various systems were devised to enable a fingerboard to produce different values of resistance, capacitance or voltage across its length, depending on the type of oscillator circuit used. On most such instruments a set of markers is employed, sometimes in the form of a 'dummy' keyboard, to indicate the finger positions required to play precise pitches.

The ability of a fingerboard to permit both discrete pitches and glissandi is not available, or only to a greatly reduced extent, on dial and space-controlled instruments. The dials consist of calibrated knobs that operate a control for the frequency of an oscillator, and are no different from those found on any analogue oscillator, whether it is a test device or part of a synthesizer. Space control normally employs the principle of hand capacitance, whereby the proximity of the performer's hand to a part of the circuit that has been brought outside the console in the form of an antenna affects the oscillator's frequency. It also describes the Saraga Generator, in which hand movements affect an oscillator's frequency by varying the amount of light that reaches a photocell. Recent systems primarily devised by musicians, in which the sound is controlled by a performer's spatial movements, include ultrasonic, holographic and infra-red detectors as well as video cameras and wind and heat sensors. In both types of controller the only method of obtaining discrete pitch intervals without the intervening glissandi is to equip the performer with a switch, push-button or variable volume control that permits the sound to be interrupted or muted during the time taken to move the dial or hand to the position for the next pitch, a process which does not permit rapid sequences of notes.

Partially polyphonic instruments. As already described, these instruments (primarily keyboard instruments, apart from later versions of the Hellertion and Trautonium which had two or more fingerboards) limit the maximum number of notes that can be sounded simultaneously. This category includes several organ-related instruments from different periods since the 1920s, a number of analogue synthesizers, especially from the mid-1970s and early 1980s, and many digital synthesizers. In each case a decision is made, based on probable repertoire, compactness and cost, to restrict the number of oscillators and thus simultaneously sounding notes to (usually) two, four, eight, 16 or 32 (on digital pianos this ranges from 12 to 128). The first instrument to do so was Harald Bode's *Warbo Formant-Orgel* (1938), for which he devised a system now known as 'assignment', expanded from that required in monophonic keyboard instruments, which assigned the four oscillators to the four highest keys depressed on the keyboard at any time; this also enabled him to give each note a different timbre. Timbre assignment to contrapuntal voices (as for Bach's organ music) was possible with instruments such as Jörg Mager's *Partiturophon* (1930-31), which had four monophonic manuals and a pedalboard. A further method of obtaining more than one timbre on a single keyboard is the 'split' keyboard, already mentioned. Small electronic organs based for economic reasons on the system of 'shared oscillators', also described above, must be considered as a special type of partially polyphonic instrument.

Polyphonic instruments. In this final category are all electronic (oscillator-based) pianos, organs (including special forms such as the electronic accordion, string synthesizer and polyphonic ensemble) and some analogue and digital synthesizers. The different numbers of oscillators used in various types of instrument has already been described, and, as with partially polyphonic instruments, is often determined by reasons of size and cost. Before the 1950s, however, an additional factor was the difficulty of designing oscillator circuits that remained in tune, so that fewer oscillators were often advisable, and the system based on frequency division ensured that at least all the octaves of any single note were precisely in tune with each other. At the same time, this also has the less satisfactory result that those intervals are not only in tune but also exactly in phase with each other, thus eliminating the complex patterns of beats that occur in polyphonic acoustic instruments, especially in pipe organs. Such problems can be effectively corrected in instruments based on a single very high frequency 'master' oscillator (usually tuned to a frequency between 1 and 4 MHz), where such 'imperfections' can be built into the circuitry. The classification of many synthesizers is less straightforward than that of other electronic instruments; a monophonic or partially polyphonic synthesizer may permit a complex sound or sequence of sounds to be triggered by each single key. In terms of the resultant sound, a better description of such instruments would be heterophonic, since monophonic, partially polyphonic or polyphonic refer primarily to the maximum number of keys that are simultaneously available.

Remote Control, Voltage Control, and MIDI

The classification system of von Hornbostel-Sachs, as applied at the beginning of this text, is primarily concerned with methods of sound generation and minimizes the importance of the activation and action stages as well as the resulting sound. In combination these latter elements define instruments into families, as is usually done in books on musical instruments and orchestration. Ignoring entirely the methods of sound generation involved, we can list the principle types of electronic instrument in terms of both familiar instruments and ones that did not exist previously. The electric guitar, electric or electronic organ and piano were, until recently, the most popular of the new instruments, while there are variants of the electronic organ that concentrate on a particular area of timbre, such as the electronic accordion and the now obsolete 'string synthesizer' and 'polyphonic ensemble' (normally featuring string, organ, piano and sometimes brass registers). A variety of smaller monophonic electronic keyboard instruments, such as those defined as 'piano attachments' (since they were often literally used in this way, although they were also played on their own in solo roles) led up to the first synthesizers; more recently (pioneered by Casio in 1980) most monophonic and some partially polyphonic keyboard instruments are aimed at amateurs and children, including the truly pocket-sized Echo Piano (mid-1980s). Those instruments that are defined as electroacoustic involve amplification, either of the acoustic principle of the instrument to which they are most closely related, or else that of an equivalent one; today MIDI makes it possible for any type of traditional instrument to control any type of synthesizer that contains the interface. Electromechanical and electronic instruments normally control oscillator frequencies by means of keyboards, with a variety of touch, 'aftertouch', velocity and other control possibilities that permit greater expressivity. Other forms of controller will be discussed in the final paragraph.

'Synthesizer' was first used to describe a musical instrument with the RCA Electronic Music Synthesizer in the 1950s, although, because of the way the term is now used, this pioneering instrument could no longer be recognised as such by most people; since they were first manufactured in the mid-1960s, all synthesizers have either been designed expressly for live performance or have been capable of it, while the RCA system was intended for studio use as what could be called a 'composition machine'. Such machines, beginning with Evgeny Sholpo's photoelectric Variofon (1932), all share with the modern synthesizer a set of controls over such elements as timbre and waveform, but differ from them in requiring at least a brief delay between the composer's defining (or, as we would now say, programming) and activating a sound and its becoming audible, thus making live performance impossible. Such machines have never been manufactured commercially, although the earliest 'modular' synthesizers (Moog, Buchla and Synket) and a number of more recent ones were initially intended for studio use rather than for live performance, and this continued with the most expensive commercial digital synthesizers like the Fairlight and Synclavier (each of which was originally developed as a studio machine). Expansion of the memory storage of 'sampled sounds' on these two systems enabled them to be introduced into recording studios as tapeless multitrack recording machines, while the largely rock-music-oriented software that has been written for microcomputers such as the various forms of the PC (IBM and others), the Apple Macintosh, Atari ST, Commodore Amiga and Acorn Archimedes does the same with MIDI-encoded materials.

The word 'control' has occurred several times in this text. It is an important clue to the nature not only of electronic instruments but also of many acoustic ones. Some historical background is necessary to place this in perspective. Up to World War II electronic instruments closely paralleled the mechanisms of the equivalent acoustic ones. The interruption to artistic activity caused by the war also served to produce unprecedented advances in all areas of electrical technology, so that several new directions were initiated in musical electronics in the late 1940s. One of these was the electronic music studio, based on a diversity of equipment intended originally for recording and broadcast studios, sound analysis, technicians' test benches, and so on. Another was a series of instruments that included more substantial methods of altering oscillator sounds, culminating in the synthesizer, including Oskar Sala's Mixturtrautonium (greatly expanded from the original Trautonium) and Hugh Le Caine's Electronic Sackbut, the first instrument to incorporate elements of 'voltage control'. Voltage control enables the varying voltage from a low-frequency oscillator to affect a selected aspect of the functioning of an oscillator or sound treatment device, replacing manual control and permitting greater speed and accuracy – in other words a form of remote control and thus of independent subsidiary activation, as described earlier. Harald Bode's modular 'signal processor' (1958-60) expanded the voltage control applications and influenced Robert Moog when he designed the series of modules that became his first synthesizer (1964). These were specifically designed to function together, unlike the motley collection of devices, often barely compatible, that made up every electronic music studio. MIDI is a more sophisticated digital control system, unifying very disparate equipment from manufacturers who had previously had no interest in such standardisation.

This shows us that any keyboard is a controller which features some degree of remote control. Instead of manually plucking the strings on the harpsichord or striking those on the piano, a remote control mechanism places the player on the other side of a keyboard, but compensates for this by providing greater flexibility. From the mid-1970s a number of rock and jazz-rock keyboard players adopted portable keyboard controllers for a synthesizer that allowed them to walk around on stage, and from the mid-1980s specialised MIDI keyboards have been marketed for remotely controlling any number of synthesizers (which may consist of 'black boxes' without keyboards). Keyboards are the ideal method of control for polyphonic instruments, but lack the subtle inflections that are possible on monophonic and partially polyphonic instruments. It took some years before other types of synthesizer controller became successful, beginning in the late 1970s with electronic percussion: special drums or drum-pads (resembling practice pads) incorporating transducers provide control information for various parameters on a synthesizer (programmable sequencer-based drum machines like the Linn Drum are special forms of digital multitrack recorders). From the same period guitar controllers such as the ARP Avatar were less effective, but since the early 1980s several models from Roland have established themselves; more sophisticated computerised controllers like the SynthAxe and Stepp DG1 had less impact. Although wind controllers had been marketed in small quantities since the mid-1970s it was not until 1987 that two major Japanese companies released models that have become more popular: Akai's trumpet-like EVI (based on the earlier Electronic Valve Instrument), and its saxophone-like derivative the EWI, and Yamaha's clarinet-like WX7 (resembling the earlier Lyricon) and subsequent newer models. Unusual controllers devised by musicians include two French systems in which laser beams are interrupted by hand movements and Michel Waisvisz's extraordinarily flexible hands.

Sources

1 This essay was originally published as 'Elektronische Instrumenten: Classificatie en Mechanismen' in the exhibition catalogue *Elektrische Muziek; drie jaar acquisitie van elektrische muziekinstrumenten* at the Gemeentemuseum (the Hague, 1988), and in a revised French translation as 'Instruments électroniques: classification et mécanismes' in *Contrechamps 11* ('Musiques Electroniques', Geneva, 1990). It is a substantially revised and expanded version of the first section of a text commissioned for publication in the exhibition catalogue *Nuova Atlantide – il continente della musica elettronica 1900-1986* by La Biennale di Venezia (Venice, 1986), under the title 'Storia ed evoluzione degli strumenti musicali elettronici'. This is its first publication in English. Much of its material is based on the author's articles (in particular 'Electronic instruments' and 'Electrophone') in *The New Grove Dictionary of Musical Instruments* (London, 1984).

Lt.-Col. Jullien, 'Applications du courant électrique, des oscillations radioélectriques et des phénomènes photoélectriques à la réalisation d'instruments de musique', *Conférences d'actualités scientifiques et industrielles* (Paris, 1929, reprinted 1930), 141.

P. Lertes, *Elektrische Musik: eine gemeinverständliche Darstellung ihrer Grundlagen, des heutigen Standes der Technik und ihre Zukunftsmöglichkeiten* (Dresden, Leipzig, 1933).

F. W. Galpin, 'The Music of Electricity: a Sketch of its Origin and Development', *Proceedings of the Musical Association* LXIV (1937-38, reprinted 1966), 71.

S. K. Lewer, *Electronic Musical Instruments* (London, 1948).

A. Douglas, *The Electronic Musical Instrument Manual: a Guide to Theory and Design* (London, 1949/1954/1957/1962/1968/1976).

W. Meyer-Eppler, *Elektrische Klangerzeugung: elektronische Musik und synthetische Sprache* (Bonn, 1949).

R. H. Dorf, *Electronic Musical Instruments*, Mineola (NY, 1954/1958/1968).

H. Le Caine, 'Electronic Music', *Proceedings of the Institute of Radio Engineers* XLIV/4 (1956), 457.

G. Anfilov, *Fizika i muzika* (Moscow, 1962/1964, English translation: 'Physics and Music', Moscow, 1966).

T. L. Rhea, *The Evolution of Electronic Musical Instruments in the United States* (Diss., George Peabody College, Nashville, TN, 1972); revised as series of articles 'Electronic Perspectives', *Contemporary Keyboard*, issues III/1-VII/6 (1977-81); reprinted in T. Darter & G. Armbruster (eds.), *The Art of Electronic Music* (New York, 1984), 4.

Contemporary Keyboard (now *Keyboard*; 1975-).

Computer Music Journal (1977-).

W. D. Kühnelt, 'Elektroakustische Musikinstrumente' in *Für Augen und Ohren* (exhibition catalogue, Berlin, 1980), 47.

D. Crombie, *The Complete Synthesizer* (London, 1982/1986).

S. Sadie (ed.), *The New Grove Dictionary of Musical Instruments* (London, 1984): articles by H. Davies and others on 'Electronic instruments', 'Computer', 'Electronic organ', 'Electrophone', 'Synthesizer', etc. and many entries on individual instruments, inventors and manufacturers.

P. Manning, *Electronic and Computer Music* (Oxford, 1985/1993).

H. Davies, 'Storia ed evoluzione degli strumenti musicali elettronici' in *Nuova Atlantide – il continente della musica elettronica 1900-1986* (exhibition catalogue, Biennale di Venezia, Venice, 1986), 17.

H. Davies, 'A History of Sampling'. *Experimental Musical Instruments* V/2 (August 1989), 17; revised in *ReR/Recommended Sourcebook 0401* ('Music under New Technology'; 1994), 5; further revised in *Feedback Papers* 40 (July 1994), 2; further revised in *Organised Sound*, I/1 (1996) 3.

C. Roads, *The Computer Music Tutorial* (Cambridge, MA/London, 1996).

Tatsuya Kobayashi

'It All Began with a Broken Organ.' The Role of Yamaha in Japan's Music Development.

Preface

Western musical instruments came to oriental countries together with Western culture and Western art they represent. Pianos and organs are basically assembled products. Therefore, as in other fields of machinery, 'reverse engineering' – taking apart, copying each component, reassembling – was done in the recipient country. The history of the Yamaha Corporation, the largest Japanese piano and organ manufacturer, provides a good example of this process, starting from a craft imitator and mass producer to their present position as a world-class piano manufacturer.

I shall start with briefly outlining the introduction of Western music and its instruments to Japan. After the Christian music, which Francisco de Xavier and others had brought to Japan, was prohibited in the late 1580s, its second arrival was the military band on board the ships of Commodore Perry which arrived in Japan in 1853 and appealed to Japanese listeners. Japanese court musicians were proud of their profession which was based on wind and string instruments brought from China and Korea from the late 6th century onwards. But they reacted favourably to Western music and regarded it as a challenge to add it to their repertoire. The records of the Department of Court Music in the Imperial Household Agency say that from 1874 onwards members of the department started to learn Western music and purchased Western music instruments. Thirty-five people were permitted to receive tuition in Western music; instruments were purchased in 1875, and the department employed John William Fenton, a British musician, as an instructor. On 3 November 1876, the birthday of the Meiji Emperor, they played Western music for the first time. Three years later they started piano lessons. One of the reasons why Western music was adopted so rapidly was the increasing necessity of performing Western music at court during reception parties for foreign ambassadors.

During the turmoil of the civil war, troops of various clans introduced a drum and fife band to provide march music. In 1871, after the civil war, the national military band followed suit. Besides such ceremonial requirements, more substantial effects were brought about by the efforts to introduce a modern educational system with Western curricula in 1872. As an educationalist with a degree from an American college, Shuji Izawa inaugurated the teaching of Western music in schools by persuading the Ministry of Education to set up a centre for music education and research; he also made the American song 'Lightly Row' into a Japanese song called 'chocho' (butterfly). The 'Music Research Institute', a nucleus of the present Tokyo National University of Fine Arts and Music, was thus founded, headed by him and guided by Luther W. Mason, an American music instructor and later by Franz Eckert, a German musician. Court musicians also joined this institute where Helmholtz's acoustics was already taught. Thus, Western music was spread in Japan parallel with Western school education.[1]

In the same year in which the Music Research Institute was established, local production of reed organs and violins started. In 1880 Torakichi Nishikawa and Sadajiro Matzunaga, musical instrument makers, copied a reed organ and a violin respectively. In 1881 Mitsunori Saita built a reed organ which served as a model for Japanese artizans and is now preserved in the Tokyo National University of Fine Arts and Music.[2]

TORAKUSU YAMAHA AND THE BEGINNING OF ORGAN MANUFACTURING

Torakusu Yamaha was born in 1851 as son of a 'samurai' in charge of astronomy in Kii clan, the present Wakayama Pref.. He later worked as a mechanic engaged in clock-making or repairing medical instruments and unsuccessfully tried to open a small business. Eventually, he moved to Hamamatsu in Shizuoka Pref. in 1887. The principal of a local primary school there asked him to repair a broken reed organ which was used in the singing course that was one of the features of the new educational system. It was probably a Mason & Hamlin harmonium imported from America and cost $ 45, equivalent to the cost of 179 gallons of rice at that time. Yamaha succeeded in repairing it and, at the same time, made a copy of it in cooperation with a metal ornament worker. The two took the harmonium to Tokyo to show it to Shuji Izawa, who taught Yamaha how to tune it and also helped him to learn acoustics in his Music Research Institute.

In 1888 Yamaha established the Yamaha Fukin (organ) Manufacturing Company, which was reorganized into Nihon Gakki (musical instrument) Manufacturing Company in 1897. Torakusu was more of a skilled entrepreneur than a talented technician. When he visited American piano manufactureres in 1899 he brought with him the components of 30 pianos including cast iron frames, while Yasuzo Nishikawa, (a son of Torakichi Nishikawa) and Shinkichi Matsumoto, competitors of Yamaha, visited the U.S.A. in 1902 for technical training.

Yamaha marketed his products through the channels of text books and musical instruments sales established by the two merchant houses Ren-ichi Shirai (Tokyo) and Sasuke Miki (Osaka). Those mercantile channels connected with schools made Yamaha's venture for mass production feasible.

Yamaha regarded himself as a pioneering entrepreneur in big business who, however, faced the problem that in the field of musical instruments the market was only small. He therefore encouraged many excellent local craftsmen like Koichi Kawai and Naokichi Ojima, but also, supported by the 'zaibatsu', embarked on large-scale multilateral management and business. This involved frequent changes of

Fig. 1: A portrait of Torakusu Yamaha

Fig. 2: Duplicate of the first organ made by T. Yamaha, 1887.

top management which led to the great labour dispute in 1926 and to the secession of the Kawai Piano Company in 1927. In a broader context entrepreneurs of this kind were common in Japan's industrial catching-up period and can be regarded as a key to economic growth.[3] In 1950 Yamaha produced about 2,600 pianos, 12,000 organs, 132,000 dozen harmonicas, and 6,000 accordions.

It should be stressed that Yamaha was lucky to be endowed with favourable locational factors. The firm originated in Hamamatsu, located halfway between Tokyo and Osaka. Apart from its moderate climate, easy communication by railroad often attracted people to live there and to launch a new business. Water and forest resources are plentiful and, above all, a remaining indigenous version of Confucianism, emphasizing industriousness and devotion to the community, had a positive effect on work ethics. It was a variety of Confucianism thought which Meizen Kimbara, a landowner and entrepreneur in the Hamamatsu area, advocated in the mid-19th century. This kind of pragmatic thinking which emphasized the pursuit of profit was common in this part of Japan and connected the region with Western thought and Western technology. Those circumstances fostered not only Yamaha, but also firms like Honda, Suzuki, and Toyota. The latter two companies had their origins in the invention of looms in the local textile industry. So, this region had for a long time been a focus of innovative talent, on which the Hamamatsu industry could build. Nowadays, Hamamatsu is the most successful Technopolis among the 29 areas which MITI has supported since the early 1980s.[4]

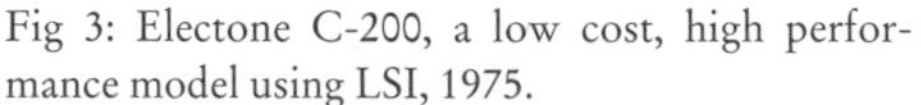
Fig 3: Electone C-200, a low cost, high performance model using LSI, 1975.

Since the late 1960s both Yamaha and its competitor, Kawai, have become successful corporations. Yamaha's Electone B-2 came on market in 1968 at a cost of 100,000 Yen. Both Yamaha and Kawai set up highly successful music courses. During the years the electone competitions which Genichi Kawakami, the president of Yamaha, organized in 1964, have become more and more international. Along with the development of IC and LSI, the synthesizer came to the scene as a complex tool controlling the rising and fading processes of the sound freely and even polyphonically.

Yamaha's business principle has always been the diversification towards a leisure-oriented complex. It comprises subsidiaries like Yamaha Motors, Yamaha Furniture, Yamaha Recreation, home electronics and semi-conductor industries. In 1987 the company revived its brand name by establishing the Yamaha Corporation as a global management system. In 1988 Yamaha produced 190,813 pianos, 145,549 electones, about 250,000 wind instruments and 3 million harmonicas, while Kawai manufactured 79,878 pianos, 12,863 electronic organs, and 30,162 electronic pianos in the same year. Yamaha and Kawai now enjoy a dominant status in the industry, outnumbering many small tenacious craft-based shops.

THE TUNER'S EYE: IMPLICIT KNOW-HOW

The job of tuning performed by piano technicians plays an important role in connecting machinery with culture. Also the art and skill of the tuner was extremely important for adapting Western musical instruments to conditions in Japan.

Musical instruments generally reflect the weather and climatic condition of the country from which they originate. For example, traditional Japanese instruments, such as *samisen, koto*, or hand drums, require a moderate extent of moisture to keep them in good condition. But the same moisture is a deadly enemy to pianos, organs or most of the Western musical instruments. This fact was noticed not only by the producers, but particularly by the tuners, who had to deal primarily with imported instruments. It was their task to select and recondition the pianos to the specific conditions of time and place. After absorbing the moisture in the damp *tatami* room, an imported Steinway piano was naturalized in Japan within a month. Tuning techniques in Japan required more subtle and elaborate care than in the Western countries. It seems that the sensibility towards nature, as expressed in Japanese art like '*waka*' and '*haiku*', was required for this new job. But those invisible skills did not play such a large role when applied to manufacturing, expecially to mass production. Company-based routine and the spread of air conditioners and substitute materials decreased the importance of tuning techniques.

Easy maintenance versus cultural quality

The old tuners had developed a feeling for the intricacies of piano mechanics and found it easy to point to the differences between a foreign and a Japanese piano. Koji Banda's, a veteran piano technicians, book, *The Musical Instrument from the West – The Example of the Piano* (1993), tells us the characteristics of Japanese pianos from the sensitive view of a musically trained technician. 'As far as Yamaha's pianos are concerned', he says, 'there is the advantage that they are very easy to tune, repair, and maintain. In this they are similar to Toyota cars. ... The most outstanding point in Japanese pianos is mechanical accuracy and precision. For example, the action parts are made very carefully and the exterior is also finished finely, but we wonder why the pianos cannot sound as beautiful as Steinways, Bösendorfers and even some Chinese pianos. The reason is that the perfectionism of the workers concentrates so much on the precision of a particular mechanism as to smother the resonance of the entire piano body.' Banda refers to the Chinese upright piano which has a tone quality similar to that of Western pianos, although the action mechanics are inferior to the Japanese ones.[5]

This veteran piano technician seems to insinuate that Japanese piano makers lack a true understanding of the heart of Western music. This impression was probably well-founded until the early 1960s, when Kawakami realized that a Steinway tone was superior. Instead of competing with this firm, he therefore planned to be associated with Steinway as their agent in Japan. When attempts to tie up with Steinway fell through in 1966, Kawakami attempted to build the world's finest concert grand piano. Three former Steinway tuners and a German piano designer employed by Yamaha dissected several Steinway grands, reproduced every minor part of them and assembled everything into a new Yamaha concert grand. Even the 'mirror' that make for good resonance was a copy from Theodore Steinway's patent of 1878.[6]

Those methods were all too common and a traditional way of technology transfer which Japanese businesses have long resorted to: 'reverse engineering'. By imitation Yamaha therefore started supplying high ranking pianists like Svatoslav Richter with a fine concert grand, made by skilled craftsmen. Although these Yamaha pianos are built by excellent craftsmen, mass production generally plays a much larger role. As hinted at above, Japanese musicians have often made efforts to absorb Western music. Seiji Ozawa, the internationally renowned conductor, was encouraged by his teacher, Hideo Saito, to master Western music. Ozawa is currently trying to perform an opera which mixes '*kabuki*' and '*noh*' elements with the 'Saito Memorial Orchestra' in Nagano Pref.[7]

THE ROLE OF YAMAHA IN PUBLIC MUSIC EDUCATION

In 1954 Yamaha started the Yamaha Music Courses in order to encourage music education for the public, mainly for children and, of course, in order to sell their instruments. The number of courses and pupils has increased from one course and 150 pupils in the first year to about 5,000 and 250,000, respectively, in 1964 and to 12,500 courses with 820,000 students in 1989. In addition to this there were courses for small children as well as guitar and electone courses in 33 overseas countries.

Fig. 4: Yamaha music course in Germany.

The purpose of this system was to create a demand for new instruments like the Yamaha Electone B-2. Both Yamaha and Kawai also adopted hire purchase systems.[8] Thus, the spread of musical instruments reflects a common Japanese style of consumption which goes hand in hand with mass production and can be compared to earlier developments in the United States. Students often hire or buy electronic musical instruments which they have become familiar with in the class room. Post-war reforms in music education which put more emphasis on instrumental performance, chorus, and music theory than on singing, extended the market for those instruments. Often young graduates from music academies, who cannot embark on a professional career as instrumentalists, find employment in those schools.

In Japan the peak of piano production was in 1975 with 393,000 pianos produced, whereas in the United States the peak was 1910 (370,000 pianos produced).

	Pianos	electronic organs
1975	317,000	262,000
1985	287,000	236,000
1994	173,000	51,000

Table 1: Output of pianos and electronic organs in Japan.

In 1995 the Hamamatsu municipality, in cooperation with the local musical instrument industries, built the Hamamatsu Museum of Musical Instruments in the Act City building in front of the Hamamatsu railway station. It houses many Western and Japanese instruments classified by sound-making principles, areas, and ages. Early Japanese-made Western instruments, a Yamaha reed organ (1907), a piano (1900), a Suzuki mandolin (1900s), military signalhorns or 'Taisho-goto' are exhibited. Those exhibits are impressive examples of materialized 'blended culture'.

CONCLUSION

The article shows that the adaption of the piano in Japan can be regarded as a typical example of technology transfer into this country which, in its early stages, was to a large extent based on 'reverse engineering'. In the case of pianos, the often overlooked 'invisible' role of the piano tuner is of vital importance. Yamaha, the best-known piano maker in Japan, made its reputation by the mass production and clever marketing of electric pianos and its system of popular music schools. But the firm has also succeeded in building individual high class concert grands which have become competitors of concert pianos manufactured by well established and highly reputed Western companies.

Notes

1 K. Horiuchi, *Ongaku Meiji Hyakunenshi* (The History of Musics in the Meiji 100 Years, Tokyo, 1968), 12-21. 32-39, & 59-60. This is a basic and orthodox book on Japanese musical history. S. Ohmori, *Nihon no Yogaku* (The Western Music in Japan, Tokyo, 1987), gives more relevant information.

2 Horiuchi, op. cit. (note 1), 58.

3 The Yamaha Corporation, *The Yamaha Century* (Hamamatsu City, 1987), 1-14, This is Yamaha's company history, highlighting the achievements of T. Yamaha and his company. R. Hiyama, *Gakki Sangyo* (Musical Instrument Industries, Tokyo, 1990), Ch. 5, 189-274, is a standard book of the musical instrument industry on the comparative basis. For a more critical view of Yamaha see Y. Satoh, *Urakaramita Yamahateikoku* (The Yamaha Empire Observed from the Back Side, Tokyo, 1996).

4 Hamamatsu Chamber of Commerce, *Enshu Kikaikinzoku Kogyo Hattenshi*, (The History of the Development of the Machinery and Metallurgy Industries in the Enshu, Hamamatsu, 1971). As to Meizen's thought K. Meizen, Marunouchi-Shuppan (Tokyo, 1968), Ch. 13. I wrote on this in my Japanese book *Gijutsuiten – Rekishikaranokosatsu, Amerika to Nihin* (Technology Transfer – Considerations from History, America and Japan, Tokyo, 1981). On recent trends in urban and regional planning and development see P. Shapira et al. (eds.), *Planning for Cities and Regions in Japan* (Liverpool, 1994).

5 Koji Banda was a Yamaha piano technician and he is now a member of the board of the Japanese Piano Technicians' Association. His *Seiyokarakita Gakki-Piano wa Kataru* (The Musical Instrument from the West – Told by Pianos, Tokyo, 1993) is one of the few books written by a piano tuner.

6 R. K. Lieberman, 'Steinway versus Yamaha' in *Steinway & Sons* (New Haven, London, 1995), 286-300.

7 S. Ozawa, K. Yamamoto, 'Is a Japanese able to conduct Western music? – My life-long experiment', *Tsusan Journal*, 1992, Vol. 25, No. 10.

8 Yamaha Corporation, op. cit. (note 3), 67-69.

Trevor Pinch & Frank Trocco

THE SOCIAL CONSTRUCTION OF THE EARLY ELECTRONIC MUSIC SYNTHESIZER

Introduction

In that we can pinpoint the location of any invention, the commercial synthesizer was designed and created in Trumansburg, NY. This is where Robert Moog had his first workshop and factory, and where he produced the legendary Moog Synthesizer in the period 1964-71. This was the first ever mass-produced synthesizer, unlike the one-of-a-kind studio synthesizers to be found earlier at the Columbia-Princeton electronic music studio, or at the Cologne music studio.[1] At about the same time that Moog produced his synthesizer, Don Buchla, in San Francisco, operating independently and unknown to Moog, also developed a synthesizer. Buchla did not produce his synthesizer commercially, mainly designing custom-built modules for studios. It is instructive to compare these two pioneering synthesizers, something we shall undertake later in the paper.

The Sociology of the Synthesizer

In this paper we apply ideas from the sociology of technology to the development of the synthesizer. The reason that the synthesizer can be examined from this analytical frame is that innovation in music, or the development of new instruments or genres of music, is like cultural innovation everywhere, including within science and technology. Formally, the early synthesizer can be described as an analogue computer, and, therefore, a machine. Thus, principles developed in the sociology of technology should have some purchase. In particular we shall draw upon ideas in the Social Construction of Technology (SCOT).[2] Essential to this analysis is the understanding of how the development of a technological artifact is negotiated between the many 'relevant social groups' who participate in its development, and the relevant social groups who share a particular meaning of the technology. Initially, there may be great 'interpretative flexibility' among the 'relevant social groups'.[3] This means there can be radically different and competing meanings of a technology. For example, in the development of the bicycle, there were many contradictory ideas among cyclists, engineers, and public interest groups as to the most appropriate design parameters. Some groups saw speed, others safety or comfort, as the most important consideration.[4]

Eventually, one meaning of an artifact stabilizes, and there is closure. In the history of the bicycle, one can see the transition from a range of bicycle designs in the

1890s, to the eventual adoption of the safety bicycle, as a process of 'interpretative flexibility' followed by 'closure'.

The last stage in SCOT is to relate the content of a technological artifact to the wider cultural milieu. The analysis of developments in musical history is especially pertinent for this approach, in that music, like everything, is a socio-cultural product.[5] It is something that is produced, re-produced, consumed, and institutionalized in a societal context. Therefore, it is our hope that this inquiry into the history of the Moog synthesizer will lead to a deeper understanding of the social environment, and historical background, of this time period, and illustrate how the synthesizer is both shared by the wider milieu and gives meaning to that milieu.

Methodological Conceit

There is very little written material, and no detailed history available of the developments we examine here. Out of necessity, we have relied on interviews with the main historical participants as our primary source material.

Fig. 1

This is an early promotional picture used by the R.A. Moog company showing a Series 900 modular synthesizer. It is being demonstrated by Bob Moog, the individual in the foreground, Frank Harris, at the back, a musician from St. Louis, and Jon Weiss, the young man sitting down, a musician who started with Moog as a work study student.

WHAT IS A VOLTAGE CONTROLLED MODULAR SYNTHESIZER?

The synthesizers designed by Moog and Buchla were much smaller than their predecessors. For instance, the RCA Mark II synthesizer was big enough to fill a room. One reason why the new machines were much smaller, and commercially viable, was because of the availability of cheap transistors in the 1960s. Moog:

> Now I can't tell you how important it was that I could buy a silicon junction transistor for 25 cents. *That's amazing* ... I can remember during that summer [working at Sperry] the technician over there undid this little box and took out thirty transistors ... These are the first silicon junction transistors. In 1957 this is $1000 for the transistors. And there they were [in 1964] for 25 cents.[6]

As well as using transistors, the new synthesizers had two important and connected design features: they were modular units and each unit was 'voltage controlled'. A modular design meant that separate units could be linked together by a patch board, rather like a telephone exchange (See figure 1). For instance, a typical synthesizer would have a number of sources of sound such as oscillators, and a number of sound processors such as filters, amplifiers, and envelope shapers.

Voltage control[7] was a technique which was evolving in electrical engineering at the time, and there was great interest in it largely stemming from the transistor. Because of the exponential relationship between input voltage and output current in a transistor, dramatic increases in current could be obtained for small voltage changes over the range of frequencies for musical pitch. The Moog synthesizer consists of a number of modular voltage control units linked by a patch board. The output of any unit can be fed as an input into another module. This means the slowly varying beat of an oscillator can be fed as a control voltage into another oscillator to provide vibrato. The early units comprised voltage controlled oscillators, amplifiers, filters, and envelope shapers.[8] Other sources of voltages such as white noise, a keyboard, or a ribbon generator[9] could be added.

In short, a voltage controlled modular synthesizer is an electronic instrument that combines simple waveforms to produce more complex sounds. Its components are built as separate and interchangeable modules that are controlled by varying the incoming voltage.

The Hobbyist Tradition

Having traced out what a synthesizer is, let us examine the route Robert Moog took toward developing such a machine. Moog's interest in electronics began in childhood. His father was an electrical engineer for Consolidated Edison in New York, and had a workshop in the basement of the family home. He and his father were avid readers of the hobbyist magazines at the time, and engaged in a number of recreational projects. Moog was always fascinated by sound, and built musical toys such as one-note organs.

An early and unusual electrical musical instrument, the Theremin, was played without being touched. The musician moved her hands in the space near its two antennas to regulate pitch and volume. It was regularly featured in magazines such as *Radio Craft* and *Radio News*, and building Theremins became a hobbyist cult of the 1940s and 1950s.

In 1949, at the age of 15, Moog built a Theremin,[10] and in 1954 he wrote an article for *Radio and Television News* on how to do it. This article elicited some interest, and he and his father were often contacted by fellow hobbyists requesting them to make Theremins. The Moogs eventually started a small business making Theremin kits. In 1961 Moog wrote a lead article for *Electronics World*, and his Theremin was featured on the cover. This helped his kit business which continued to thrive. When Moog went through graduate school in engineering physics at Cornell he sold many Theremin kits, at $50 each.[11]

A Bumper Crop

Moog's Theremin kits were sold in New York City by a sales representative, Walter Sear, who made his living playing the tuba in Broadway shows. Sear also sold tubas. In December 1963, Sear was showing his tubas at a regional educator's convention, and he invited Moog along to demonstrate the Theremin. It was at that exhibit that New York experimental musician Herb Deutsch introduced himself to Moog. Deutsch had been using a Moog Theremin and was delighted to meet the designer in person.

Deutsch invited Moog to a concert of his music in a loft belonging to Jason Seley, famous for his sculptures made out of automobile bumpers.[12] The music was electronic, including a prerecorded tape accompanied by a percussionist who hit Seley's sculptures. Moog: 'I was excited. What Herb was doing artistically just, you know, it sounded like fun. It sounded like something that I'd like to help him with.'[13]

What is that Weird Sxxt?

At this point, Moog had a small shop in Trumansburg, NY, north of Ithaca, employing two people designing musical instrument amplifier kits. Deutsch came for a visit in the late Spring of 1964. Moog:

> And I asked him, 'You know what do you want to be able to do Herb?' He says, 'Well, I want to make these sounds that go brooo, brooo broough' [yelping sound] . . . And somehow, during those first months of 1964 when I was thinking of what to put together for Herb, voltage control suggested itself. I knew of voltage control. Voltage control was a technique that was just becoming practical because of the properties of these new silicon transistors that were coming out.[14]

Moog put together a breadboard[15] with three circuits: two voltage controlled oscillators and a voltage controlled amplifier. Here is what happened next:

> Herb, when he saw these things, sorta went through the roof. I mean, he took this and he went down in the basement where we had a little table set up and he started putting music together. Then it was my turn for my head to blow. I still remember, the door was open, we didn't have air conditioning or anything like that, it was late Spring and people would walk by, you know, they'd listen and they'd shake their heads. You know, they'd listen again – what is this weird shit coming out of the basement?[16]

These were the first sounds of the first Moog synthesizer.

THE ROLLER COASTER GETS ROLLING

At this point, Moog regarded the project as something which was 'really neat',[17] but he didn't immediately abandon his amplifier kits. As he told us, 'This sorta stuff I did for the hell of it while everybody else tried to make the shop go ... '[18]

Moog and Deutsch demonstrated the early modules (consisting of two Voltage Controlled Oscillators and two Voltage Controlled Amplifiers) to the Toronto Electronic Music studio. There, composers Myron Schaffer, Anthony Gnazzo, and Gustav Ciamaga got very excited. Moog: 'Myron Schaffer encouraged me. He says ›Boy, this is great. I'm sure a lot of people will be interested in this‹. Little by little the idea began to form that maybe we could sell this shit.'[19]

The endorsement of the Toronto studio was important because Deutsch at that time was a young academic composer, working with limited resources, unlike the composers associated with the Toronto studio who were well-funded by the Canadian National Research Council.

From there on Moog started to take orders, including one from Vladimir Ussachavesky, of the well-known Columbia-Princeton Studio for Electronic Music, who wanted particular modules built. Ussachavesky was provided with a voltage controlled envelope generator, which in turn led Moog to develop ADSR (Attack, Decay, Sustain, Release) terminology,[20] which subsequently became the standard for specifying envelopes on all synthesizers.

During this period Moog saw himself as a builder of equipment for electronic music composers, including the electronics for tape recorders. In 1967, after three years selling customized units designed for individuals, Moog decided that the units could be sold as complete standard systems. This was the first time that his catalogue used the word synthesizer. At about $10,000 each, these 'synthesizers' were mainly for use in studios or for individual wealthy musicians.

THE MAN FROM MOOG AND HIS FABULOUS 'SANITIZER'

There is an early promotional picture used by the R.A. Moog company (See figure 1). It shows a Series 900 modular synthesizer, apparently being played by three people. This photograph is a staged mock-up. Two of the people are musicians and one is an engineer. Bob Moog, the engineer, is the individual in the foreground.[21] The person at the back is Frank Harris, a musician from St. Louis, who was an early customer of Moog's, and who happened to be in the studio that day. The young man sitting down is Jon Weiss, a musician who started with Moog as a work study student, and who proved adept at learning the synthesizer. Moog counted on him for insights into the sort of technology musicians needed.[22] Because he was the person who delivered the synthesizers and taught their new musician-owners how to use them, Weiss became known as the 'Man from Moog'.

There is something striking about this picture. It has to do with the left hands of the two musicians. Both have their left hands extended upwards adjusting potentiometer knobs. The right hands are lower – in the case of Weiss, his hand is on the keyboard. This was a posture deliberately used by Moog in his advertising. As he told us, this seemed to encapsulate the link between the music and the machine:

> The keyboards were always there, and whenever someone wanted to take a picture, for some reason or other it looks good if you're playing a keyboard. People understand that then you're making music. You know [without it] you could be tuning in Russia! This pose here [acts out the pose], graphically ties in the music and the technology.[23]

The visual representation of the synthesizer to be used in promotional literature must somehow capture the fact that the machine plays music, but that it is not just a keyboard. The synthesizer is an odd instrument because, unlike most instruments, it is very hard to play a tune on it, and, unless it is outfitted with a speaker or headphones, it doesn't make a sound. Weiss:

> People don't understand ... that the synthesizer itself doesn't produce any sound. When I got the synthesizer to London [for Mick Jagger's use in filming the movie 'Performance'] and I brought it onto the set ... I had to go through this with the English workers saying, 'Agh, it's a fabulous sanitizer, and what does it do?' you know, 'Play us a tune'. ... Moog heard that so much that in one series of synthesizers he put a little speaker and amplifier in one so that you could actually hear something. People couldn't conceive that this is an instrument, but it doesn't do anything.[24]

Playing the 'sanitizer', required musicians to learn a whole new form of 'body technique' (to use Marcel Mauss' term).[25] The use of the synthesizer for live musical performance, a rarity in those early days as the time needed to set up the patches made it more suitable for studio use, involved a whole new series of practices. The patch changes would take so long that one early synthesizer band, Mother Mallard, from Ithaca, NY, played cartoons between pieces to keep the audience amused.[26]

The Keyboard Becomes a Source of Certainty

In terms of the development of the synthesizer, something even more interesting is going on in this picture. Why use a keyboard at all? Indeed, Moog initially saw the keyboard as only one option for a controller. He developed a new controller, the 'ribbon controller', a long potentiometer strip which worked by sliding a finger up and down its surface. Don Buchla, who was an experimental musician as well as synthesizer designer, was even more radical. Buchla rejected the use of a conventional keyboard.[27] For Buchla, it was restrictive to use an old technology associated with wires and hammers with the new electronic source of sound. He wanted to 'build an *intentional* electronic musical instrument.' Buchla:

> No, I think you have to go a little further back and say what does electronic music mean? The term bothers me a little bit, even now. To me it meant simply the source of sound was electronic rather than vibrating strings, membranes, and columns of wind. And to me, that meant that it was a potentially new source, and therefore instruments based on it would probably be new and different. I saw no reason to borrow from a keyboard which is a device invented to throw hammers at strings, later on operating switches for electronic organs, and so on. But I didn't particularly want to borrow the keyboard to control. ... I tried once to put a keyboard on my system ... And I found myself overwhelmed by the psychological aspect of looking at this very familiar twelve tone structure, and wanting to do music that was very much against what I was conditioned to do. ... If you ask the man on the street, 'What's a synthesizer?' He will reply, 'A synthesizer is a keyboard instrument.' If you go into a retail store and say, 'I want to see some electronic instruments,' they'll send you to the keyboard department, because a synthesizer is a keyboard instrument by default.

The tying of the keyboard, the wedding of the keyboard with the synthesizer was a disaster for a creative composer who doesn't want to do twelve-tone melodic, and doesn't want to imitate violins and saxophones and so on and so forth, who wants new kinds of dynamics, and networks attached to his sounds ...[28] Weiss:

> He [Buchla] had a distaste of the keyboard, and I think for a legitimate reason. In that he didn't want his machine to be a glorified electric organ. So the only controller that he provided like that was a touch-sensitive pressure pad ... His designs were wild and wonderful. Moog's were conservative, rigorous, and well-controlled ... all you had to do was look at the names of some of his modules. Moog on the 900 Series had VCO, VCF, envelope generator, blah, blah. Buchla has 'Source of Uncertainty', and sample and hold circuits, and bizarre things that he designed while he was high.[29]

Buchla's machine had much in common with Moog's. Both were modular voltage controlled synthesizers, but they were critically different in how they were to be used. The interesting issue here is that a new sort of instrument is coming into being, and radically different meanings are being given to it. There is 'interpretative flexibility.'

It is worth considering how few new musical instruments ever become commercially viable and mass-produced. If a new instrument does come along, how do people recognize that instrument and its sound, and how does it get incorporated into the wider corpus of musical culture? In the history of the synthesizer, there is a path that can be traced from the Moog, through the Mini-Moog (a hard-wired Moog with no patching), to the Yamaha DX7 (the first commercially successful digital synthesizer), and all the myriad Casio keyboard synthesizers. In this developmental path, the synthesizer finally becomes a version of the electronic organ, only with a greater range of voices and special effects. The first step in this process is the idea that the keyboard is the preferred interface, and the notion that control voltages are correlated with conventional octaves.[30]

It is at this point that the 'relevant social groups', of musicians and synthesizer users, voice their preference, through sales and personal feedback, to Moog. The original interpretative flexibility, where many controller interfaces were imagined and possible,[31] has collapsed into only one repetitive configuration, the keyboard.

The paradox of the Moog becoming defined as a keyboard instrument was apparent to Moog. Unlike Buchla, who worked alone and who wanted to make unique, individual instruments, Moog came from a tradition of mass-production. Moog's dream was to have a small shop that produced electronic devices. It was no particular concern of his how musicians used his instruments, he was simply providing them with tools. He told us that he could just as easily have been designing power drills as synthesizers.[32]

It is important not to view Moog's attitude simplistically, as the work of a short-sighted engineer. Moog's view was that if the musicians wanted keyboards he would supply them as another module. This is an indication of Moog's responsiveness to the needs of the musicians with whom he was collaborating. Unlike Buchla, who was himself a musician and held on to a philosophical ideal even if it meant that his instruments would remain inaccessible to the majority of musicians, Moog was involved in developing an instrument that could be purchased and utilized by every musician. Moog:

> It began strictly as separate parts ... nothing biasing one way of hooking up versus another way of hooking up. And we just responded to demand. I didn't stand there and say, you know, 'I'm going to do it the right way, and you can't have it any other way ... '[33]

Compare this to Buchla:

> I'm an instrument builder ... I don't build machines, I never have built a machine. I build only things that you play. I don't build things that you program, I don't build things that are involved in issues. I am an old fashioned builder of instruments.
> And it's a niche market. I'm not interested in serving the commercial markets, even if people beat a path to my doorway. I run the other way. I'm just not interested in the directions the music has taken. I'm not interested in it personally, and I'm not interested in serving them.[34]

Switched on Bach

The difference between Moog and Buchla, and the meaning given to their respective instruments, can also be seen by the type of music that gets played on synthesizers. We probably never would have heard of Moog if it had not been for the recording sensation of 1969 – Wendy Carlos' 'Switched on Bach'.[35] On the album cover a wigged figure (presumably Bach) is pictured listening to a Moog synthesizer. This record became a best seller, and made Carlos and Moog famous. It is the best selling Bach record of all time, and one of the best selling classical music records ever.[36] Moog:

> The conventional wisdom back in the music business ... [was that] nobody believed that this kind of thing could be used as anything more than a novelty. You couldn't make real music with it. You couldn't be expressive with it. You couldn't make it swing. And then Carlos and a few other people demonstrated that they were wrong. You know, they just did an end run around the music business. And *then* in 1969 all hell broke loose. Everybody had to have a, you know, every commercial musician had to have a synthesizer.
>
> In 1968 at the AES [Audio Engineering Society] at the Electronic Music Session, and it was a big session, I gave a paper ... this was a couple of weeks before 'Switched On Bach' was to be released. And I had probably 100-150 audio-engineers listening. And Carlos let me play a track from 'Switched On Bach' as an example ... I put the tape on and just walked off the stage. And I can remember people's mouths dropping open, and I swear I could see a couple of those cynical old bastards starting to cry ... Those cynical experienced New York ... [engineers], they had their minds blown ...[37]

The success of the record got Moog onto prime time TV, into *Time* magazine,[38] and the *New York Times*. It also prompted a deluge of orders, as every musician and recording industry hack sent their people out to buy a Moog to try and emulate Carlos. None ever did it successfully.[39]

Walter Sear, who by this time had become Moog's synthesizer representative in New York, sold forty module synthesizers to commercial music producers in New York in 1969. His representative on the West coast, Paul Beaver, sold even more. 1969 was the only year that R.A. Moog of Trumansburg recorded a profit. Weiss:

> I could see the difference, and there was a world of difference pre-'Switched on Bach', and post-'Switched On Bach'. Before 'Switched On Bach' came out, the synthesizer was basically resigned to well to do academic institutions, a few private individuals, very few ... And it was pretty much considered lunatic fringe, there's no question about it, you know, weird space sounds ... there was some rigid thinking about what's music and what isn't music, what's permissible and what's not, and then Carlos came along. Like Wham, and then suddenly the world thought, 'Oh yeah, this is great ... '[40]

The use of the synthesizer to play conventional keyboard music, although with new timbres, further defined the Moog synthesizer as an instrument to be interpreted within the traditional genre of music. It may have been switched on – but it was still Bach. Curiously, when Moog went on the 'Today' show with his synthesizer, switching it on actually became an issue. There was a dispute between the electricians' union and the musicians' union as to who should switch on and operate the synthesizer.[41] Was it an electrical device or a musical instrument? Subsequently, there were many worries from the musicians' union that the synthesizer might cost them jobs.

Pop musicians and the newly emerging rock bands also took up the instrument with a vengeance. The Beatles used it on their album 'Abbey Road'. One of the popular songs from that era, George Harrison's 'Here Comes the Sun', utilizes the spacy melodious tones of Moog, and the song 'Maxwell's Silver Hammer' uses the synthesizer to emulate a conventional instrument, the French Horn.[42] Keyboard 'whizzes' became associated with the synthesizer, people such as Keith Emerson and Rick Wakeman in the UK, and Don Preston in the US. One of the most familiar pop uses of the synthesizer was the closing bars to Emerson, Lake, and Palmer's, 'Lucky Man'.[43]

For both pop music and so-called serious music, the Moog synthesizer was increasingly being taken up as a keyboard instrument. The new-found success of the keyboard players with the instrument further helped define the meaning of the synthesizer. Obviously, in the modern recording industry, the success of records plays a key role in defining how sounds are heard and in recognizing the capabilities of instruments.

The sound of the Moog and the sort of music that gets played on it can be contrasted with the sound of the Buchla. At the same time that 'Switched on Bach' was released, electronic music composer Mort Subotnick released 'Silver Apples of the Moon',[44] played on a Buchla. The weird, clearly electronic sounds on this recording represent a completely different genre of music and sound than 'Switched on Bach'.

The use of the Moog to play keyboard music in this way wasn't necessarily followed by all musicians. For instance, Weiss, although shown in the staged picture playing the keyboard, always used the instrument in his own compositions without the keyboard. Weiss:

> I was interested in the synthesizer because it gave the potential of creating sounds that couldn't be produced by any other means. And create kinds of sounds, kinds of music, that you just simply couldn't orchestrate ... The beauty of it was you could conceive of a sound, or sound event, in your mind, and you wouldn't have a chance in hell of getting an orchestra to be able to create it ... When I used the synthesizer I didn't use the keyboard ever. That's a ... confusion that many people have a difficult time overcoming. They feel that a synthesizer is a keyboard instrument ... I had no interest in using the synthesizer to create instrumental sounds ... if you wanted something to sound like a French Horn then play a French Horn. Why use this machine to do just that?[45]

We do not want to suggest a deterministic history of the synthesizer. That Weiss could use the Moog in this way reminds us again of the interpretative flexibility which resided in the synthesizer at this early stage of development. Other possibilities of construction and design always remain open on the margins (e.g., Buchla, although he experimented with keyboards, developed his own distinct controllers). We are describing the main commercial path of development. Since designs can always be reinterpreted and reconfigured, closure is never final.

The Moog synthesizer, like the Theremin before it, was quickly taken up by Hollywood. One obvious usage was its capabilities for producing sounds suitable for science fiction or horror movies.[46] Dave Borden, the musical director of Mother Mallard, told us how he was commissioned to do some sounds for the movie 'The Exorcist'. 'I sent him three things and he used them all in the film ... He paid me well for three minutes. It took me about one morning, he paid me a few thousand dollars ... '[47]

The technical versatility and flexibility of the Moog was increasingly being exploited, and significantly, it was finally able to make the sound of music.

Swurpledeewurpledeezeech![48]

The meaning given to the synthesizer as an instrument is inseparable from the kind of sound it produced. Here it is important to explore the difference between the machine as an abstract entity – as an early form of a universal music machine capable of producing any sound – and its material form within the practices of users and listeners. There is no doubt that the early synthesizers had a particular sound and this was related to the technology.[49] Several people told us that the distinctive sound of the Moog came from its voltage controlled filter. This filter was a unique design, and was the only element that Moog patented, although this did not prevent others from copying it.[50]

The peculiarities of these early instruments were well known to musicians. There is a story told by Brian Eno of how his synthesizer (a VCS-3) developed a fault with a unit called the ring modulator. Eno loved the sound this fault produced, so every time he had his synthesizer serviced he attached a note to it telling the technician to leave the ring modulator alone.[51]

A similar story is told by Weiss about the musician Sun Ra, and his 'Solar Arkestra', one of the most interesting characters to visit the Trumansburg shop. He always travelled with a huge entourage in a fleet of 20-year-old black Cadillacs:

> He came to Trumansburg back in 1968. This was a fairly rigid sleepy little New York state town, and here's this bizarre looking Black guy in his robes and stuff sitting down in the local ice cream parlor ... It was in the days of the Mini-Moog; he saw one and thought that he wanted to incorporate it into his act ...
>
> I happened to hear this machine, and he had taken this synthesizer, and I don't know what he had done to it, but he made sounds like you had never heard in your life. I mean, just total inharmonic distortion all over the place, oscillators weren't oscillating any more, nothing was working, but it was fabulous. He had taken that machine and somehow, I know he hadn't gotten inside of it; who knows what he had done or what it had been subjected to, but he created these absolutely out-of-this-world sounds, that the engineers could never have anticipated. That's the important thing about all this, you know, that's analogue and this is even before integrated circuits, there was a lot of instability. Moog couldn't put his finger on it ... So it really was an instrument more than a machine, we've been talking about it as a machine; it took on characteristics of whoever was using it.[52]

The inaccuracies of the instrument were something noted by Borden[53] and Weiss:

> The interesting thing about the synthesizer is that the machine was not as accurate as all the engineers wanted it to be. And the wonderful thing about that is, there were characteristics of the Moog synthesizer that existed only because of certain inaccuracies in the equipment that resulted in wonderful and bizarre events, that were many times positive. Jim Scott [a Moog engineer] and I were talking about this just the other day, and Moog knew that himself. Here was a machine that was supposed to be absolutely infallible, you can't get anything more well-defined

> than a perfect sine wave, but it wasn't perfect. The oscillators wouldn't keep exactly perfect pitch, there would be some drifting, depending on the temperature, depending on inaccuracies in the line voltage they would change. They were temperamental. One day you'd get something incredible, and you'd try and get it the next day and you couldn't. Your tape recorder was your best friend. If you got something, some incredibly complex sound, and you worked on it and you finally got it then you'd better get it down on tape because you'd probably never get it again. That was what was so wonderful about the machine, in the sense that it was an instrument, it wasn't machine. A machine would have created no inaccuracies, and I think that's why these computer digital generated sounds are not as interesting as the analogue sounds. Because they are too accurate ...[54]

There is a paradox here that has created an interesting tension. Beginning with Moog, engineers have worked vigorously for years developing the synthesizer to be more and more closely able to mimic the sounds of conventional musical instruments. Eventually, they were able to produce synthesizers that, for most purposes, were indistinguishable from conventional instruments. Interestingly, the closer that the synthesizer came to this mimicry, that is, the more predictable and machine-like it became, the less interesting it was to many avant-garde musicians. It became just another piano. The imprecision of the old instruments was eventually lost as the machines developed technologically, and Buchla's fear of the keyboard synthesizer becoming a 'a disaster for the creative composer', may have been realized.

The imprecision was finessed out of the synthesizer by the engineers who recognized that making a more precise instrument was a fascinating puzzle (and who had, as time passed, more advanced components), and the musicians who wanted to use it to emulate other instruments. This was not a determined outcome for the synthesizer, but the influence of the relevant social groups of engineers and musicians. The result was mixed: it worked for some musicians, but was lacking for others who preferred the analogue imprecision.

Synthesizer Daze

No story of the early synthesizer days would be complete without some reference to the wider culture of which the synthesizer was part, indeed an often over-looked part. The story of the synthesizer is not one of 'technological pessimism' – for many people the technology was an integral part of the liberating late 1960s and early 1970s.

It is difficult to conceive of now, but during this period music was integrated into something much larger. The synthesizer, and the sound it produced, was part of the counter-culture, the 'sixties thing', and the psychedelic revolution. Life styles blended with the technology, communal living and mind expansion were the order of the day, and music was often made, and listened to, in a mind-altered state. This is how people participated in the early Moog (and John Cage) concerts. Weiss described what was billed as the first ever live electronic music concert, held in the garden of the Museum of Modern Art in New York:

> These machines came down the day of the concert, and the machines weren't ready. They were there but not working. We got them to the garden and Moog and his engineers were there with soldering irons up to the minute before the whole thing started. And it was wild ... There was music, and it was cranking and it was cranking, and then the power went out. But nobody cared. It was in the garden at night, the garden at the museum was filled, it was outdoors and everyone cheered. It was like a happening ...[56]

Experiencing the music was a form of transcendence, a way to mind expansion – especially the hypnotic minimalist style of bands such as Mother Mallard. Droves of people would show up and lie around stoned on the floor as the band played. The ability to use the synthesizer, and make such music, was also an expansive experience for many musicians. Linda Fisher, a New York artist and musician in Mother Mallard, described this way of experiencing the music in the course of a discussion as to whether the synthesizer is gendered:

> My experience, whatever it was, whether it came from being a woman or just from being Linda, I'm sure most of it came from just being Linda, to be able to express that reality in sound gave me palettes that, you know, were really unimaginable to me before that ... Living through that time was very intense politically, but there was also that sense that we could do anything we wanted. It was very idealist that we could be who we wanted, of course, and always be totally politically responsible, and totally creative, everything, there really was that feeling in the early seventies.[57]

Weiss also describes a similar impression:

> There was a real interest, there was an interest in expanding consciousness, which meant expanding musical palette, and, you know, creating environments and sound ... I think we all, I think everyone was plotting for a brave new world. Everything would be a little nicer, everything would be a little more harmonious ...[58]

Moog himself used to introduce Mother Mallard and attend many 'happenings', plus go to recording studios with musicians. But he was never part of the culture, or the music of the time: 'Back then I didn't have a hell of a lot of time to listen to it. I hardly knew who the Beatles were, I didn't know who the Grateful Dead was. You know, I learned about them over the years, but I just didn't sit around listening to music.'

There is no doubt that the 'spacy' sounds the synthesizer could produce perfectly matched people's explorations into inner-space. With bands such as Hawkwind and the Pink Floyd in Britain, and Jefferson Airplane in the US, who packaged their material with the imagery of outer space exploration, the synthesizer provided a means to experience all sorts of spaces.

CONCLUSION: MIRACLES OF SOUND

The story of the analogue synthesizer is like that of many technologies. One meaning stabilizes, and the other meanings slowly vanish or play a smaller role within niche markets. Also, as with many stories of technological innovation, the pioneers are often not the ones to receive the financial rewards. Although the Mini-Moog was a commercial success, the market for the modular units collapsed in 1970 after the recording industry became disillusioned with the Carlos imitators. Moog never went bankrupt, but in 1971 was forced to sell his business and his name.

The meaning of the synthesizer as a keyboard instrument, a meaning which was slowly embedded in the technology, reinforced in the way the instrument was used and the type of music performed on it, and which in turn was responded to by changes in the technology (e.g., Mini-Moog), in the end wins out. The pay day only comes when Japanese companies such as Yamaha enter the market in the early 1980s, and every pop star had to have a keyboard synthesizer. Moog is by then out of the business, as were many of the early companies, missing out on what was, by 1990, a three-billion dollar a year business in the US alone. Today, Moog manufactures Theremin kits in North Carolina.

We have shown how an inventor/engineer, in collaboration with a relevant social group of musicians, initiated the idea for, and fashioned, a new kind of musical instrument. The Moog synthesizer allowed the musician to produce a previously unheard category of sounds. From a flexible variety of possible control configurations, the synthesizer eventually stabilized into a keyboard instrument, widely accepted by pop and rock musicians, composers, and creators of original sounds. As the keyboard synthesizer became established, a continuing design effort was made to make it predictable, easy to use in live performance, and to enable it to replicate the sound of conventional instruments. This allowed it to become widely available. The synthesizer had evolved into the instrument that everyone thought they wanted, but, as it lost its early promiscuity, it also lost some of its original appeal to instrumentalists, composers, and engineers on the frontiers of musical creativity.

The synthesizer's stabilization as a keyboard instrument, while an attempt to increase its versatility, and a major step in allowing wide distribution and maintaining its commercial viability, may have begun the process of delimiting its creative freedom. The synthesizer's musical identity, today almost indistinguishable as unusual to a new generation of listeners, still remains symbolic of the initial dream for an innovative instrument that could perform miracles of sound. The synthesizer, once an instrument upon which the radical musician could manufacture any sound he or she could imagine, has stabilized, perhaps where it originally began, as a magnificent music machine.

Notes

This is a revised version of an article published in *ICON. Journal of the International Committee for the History of Technology* 1998, 4:9-31.

1 For the early history of electronic music see T. Darter and G. Armbruster (eds.), *The Art of Electronic Music* (New York, 1984). For an interesting account of the early days of the Cologne studio, see E. Ungeheuer, 'Concepts of Technology in the Early Days of Electronic Music',

paper presented to the 23rd Symposium of the International Committee for the History of Technology, Budapest, Hungary, August 7-11, 1996.

2 W. Bijker, T. Hughes and T. Pinch, *The Social Construction of Technological Systems: New Developments in the History and Sociology of Technology* (Cambridge MA., 1987).

3 Bijker, Hughes, Pinch, op. cit. (note 2).

4 Bijker, Hughes, Pinch, op. cit. (note 2).

5 Our forays into the sociology of music have thus far been rather limited. There is undoubtedly a large strand of work in ethnomusicology which is relevant to this project. For some work in the sociology of music which we have found informative see S. Frith and A. Goodwin (eds.), *On Record: Rock, Pop and the Written Word* (London, 1990); S. Frith (ed.), *Facing the Music: Essays on Pop, Rock and Culture* (New York, 1988); S. Frith, *The Sociology of Rock* (London, 1978); M. Chanan, *Repeated Takes: A Short History of Recording and its Effects on Music* (London, 1995); P. Wicke, *Rock Music: Culture, Aesthetics and Sociology* (Cambridge, 1993); P. Friedlander, *Rock and Roll: A Social History* (Boulder CO., 1996).

6 Interview with Robert Moog, June 5, 1996.

7 R.A. Moog, 'Voltage-Controlled Electronic Music Modules', *Journal of the Audio Engineering Society*, July 1965, No. 3, 13: 200-6.

8 A control voltage which can be applied to any voltage-controlled parameter to give shape to a note.

9 A slide device for producing variable resistance.

10 A. V. Glinsky, *The Theremin in the Emergence of Electronic Music* (Ph.D.: Thesis, New York University, 1992). See also the recent movie documenting the story of the Theremin. 'Theremin' (directed by Steve Martin, Orion, Hollywood, 1995).

11 Moog's graduate research at Cornell was in low-temperature physics and had little direct connection with his pioneering development of the synthesizer.

12 Seley became head of the Department of Fine Arts at Cornell and some of his sculptures of automobile fenders can still be found on campus.

13 Interview Moog, op. cit. (note 6).

14 Interview Moog, op. cit. (note 6).

15 A board on which an experimental model or prototype is constructed.

16 Interview Moog, op. cit. (note 6).

17 Interview Moog, op. cit. (note 6).

18 Interview Moog, op. cit. (note 6).

19 Interview Moog, op. cit. (note 6).

20 The four controls found on the commonest type of envelope generator, which determine the relative shapes of the segments of the envelope.

21 Although he learnt piano as a kid, and likes and is knowledgeable about music, Moog does not regard himself as a musician.

22 Dave Borden, another Ithaca-based musician, also played a similar role as resident composer.

23 Interview Moog, op. cit. (note 6).

24 Interview with Jon Weiss, May 1996.

25 See M. Mauss, 'Les Techniques du Corps', *Journal de Psychologie Normale et Pathologique*, 1934, 32: 271-93.

26 Interview with David Borden, May 3, 1996.

27 See Jim Aikin's *Keyboard* interview with Buchla, reprinted in *The Art of Electronic Music*. For the early history of electronic music, see *The Art of Electronic Music* (note 1).

28 Interview with Don Buchla, May 27, 1996. Interview in *Keyboard*; in Darter, Armbruster, op. cit. (note 1). For the early history of electronic music, see Darter, Armbruster, op. cit. (note 1). See also M. Vail, 'Buchla's First Modular System: Still Going Strong After 30 Years', *Keyboard*, October 1992, 45-6, 48, 50.

29 Interview with Jon Weiss, June 1996.

30 It must also be added, however, that in one way Buchla's design is more conservative than Moog's. In the Buchla system, control voltages and signals are kept separate and designated by different wires and plugs. This means that with a Buchla it is hard to use an inappropriate signal voltage as a control. The Moog allowed for more experimentation in this respect and enabled some very unusual sounds to be made.

31 Instead of a keyboard, Buchla's controller was called the 'Kinesthetic Input Port', a device for sensing movement.

32 Interview Moog, op. cit. (note 6).

33 Interview Moog, op. cit. (note 6).

34 Interview with Don Buchla, May 27, 1996.

35 W. Carlos, 'Switched on Bach'. *CBS Records*,1968.

36 I. Berger, 'The Switched on Bach Story', *Saturday Review*, January 25,1969, 45-47. H. Saal, 'Electric Bach', *Newsweek*, February 3, 1969.

37 Interview Moog, op. cit. (note 6).

38 'Into Our Lives with Moog', *Time*, March 7, 1969, 50-51.

39 One exception may be the film 'Last Escape from New York' by John Carpenter (1981), the music for which was produced on a synthesizer. This film contains a superb rendering of Debussy's La cathédrale engloutie.

40 Interview with Jon Weiss, May 1996.

41 Interview with Dave Borden, May 3, 1996.

42 George Harrison purchased a Moog Synthesizer in Los Angeles in November 1968. Harrison made an album of electronic music, 'Electronic Sounds', issued by the Apple experimental label 'Zapple'. On 'Abbey Road' John Lennon also uses the synthesizer on the track 'I want you'. Bob Moog told us that he thought he detected the sound of his synthesizer on an earlier Beatles track, 'Lucy in the Sky with Diamonds' on the album 'Sergeant Pepper's Lonely Heart Clubs Band'. However, in discographies of the Beatles' music (e.g. I. Mac Donald, *Revolution in the Head: The Beatles' Records and the Sixties* (London, 1994) there is no reference to a synthesizer being used on this track.

43 Emerson was a rare exception in taking the Moog modular synthesizer on live tour. Most musicians favored the mini-Moog for live performance. For a rather more interesting use of the mini-Moog by a rock band, 'see' Don Preston's playing on various Frank Zappa and the Mothers of Invention albums (e.g. 'Lonesome Electric Turkey' on Live at Fillmore East). For Frank Zappa and a series of reflections on modernism, postmodernism and music, see B. Watson, *Frank Zappa – The Negative Dialectics of Poodle Play* (London, 1995).

44 M. Subotnick. 'Silver Apples of the Moon'. Nonesuch Records,1967.

45 Interview with Jon Weiss, May 1996.

46 This fits with Ben Watson's point that the avant garde is only taken up by mainstream culture in horror movies.

47 Interview with Dave Borden, May 3, 1996. The 'him' was the director Billy Friedkin.

48 'Swurpledeewurpledeezeech!' *Time*. November 4, 1966: 44.

49 For an exploration of the issues raised by trying to emulate conventional instruments see T. Pinch, 'Towards a Sociology of the Electronic Music Synthesizer: Some Ideas', paper given at CRICT, Brunel University, February 8, 1995.

50 Moog, op. cit. (note 7).

51 Interview with Brian Eno, 'Keyboard' in Art of Electronic Music (note 1).

52 Interview with Jon Weiss, May 1996.

53 Interview with Dave Borden May 3, 1996.

54 Interview with Jon Weiss, June 1996.

55 Y. Ezrahi, E. Mendelsohn and H. Segal, *Technology, Pessimism, and Postmodernism* (Boston, 1994).

56 Ezrahi, Mendelsohn, Segal, op. cit. (note 55).

57 Interview with Linda Fisher, May 14, 1996.

58 Interview with Jon Weiss, June 1996.

JÜRGEN HOCKER

MY SOUL IS IN THE MACHINE – CONLON NANCARROW – COMPOSER FOR PLAYER PIANO – PRECURSOR OF COMPUTER MUSIC

INTRODUCTION

As long as musical instruments have existed, artisans and musicians have tried to construct instruments which play by themselves. Descriptions of self-playing instruments are to be found in Ancient Greece (Heron's self-playing organ pipes in Alexandria) and in Arabia (the Mûsà brothers' automatic flute-player from Bagdad). The oldest surviving mechanical musical instruments are 'Glockenspiele' (carillons) of the late Middle Ages. Many composers also discovered the charm of these instruments very early and composed pieces for them; Händel, C.P.E. Bach, Haydn, Mozart and Beethoven composed for small self-playing organs which were usually built into the cases of grandfather clocks (so called 'flute clocks'). At the beginning of the 19th century the 'music machinist' Johann Nepomuk Mälzel, famous for, but falsely credited with, the invention of the metronome, even created a whole self-playing orchestra, the 'Panharmonikon', and Ludwig van Beethoven composed his 'Wellingtons Sieg oder die Schlacht bei Vittoria' op. 91 for this instrument.

During the 19th century, self-playing instruments were largely ignored by composers. It was not until the beginning of the 20th century that interest in them was reawakened as a result of a pioneering invention: the player piano. In 1904, the firm of Welte in Freiburg, Germany, succeeded in recording the original performance of a pianist on perforated paper and reproducing it with all dynamic and agogic details. Many composers and pianists – including several pupils of Liszt – used this procedure in order to make their piano-playing accessible to a wide spectrum of the population; at a time when it was not yet possible to record the sound of a piano on a gramophone disc satisfactorily. Thus, perforated music rolls have survived with interpretations by d'Albert, Busoni, Paderewski, Godowsky, Grieg, Reger, Debussy, Ravel and others. Even the brilliant technique of the young Horowitz can thus be appreciated.

In the 1920s the firm of Welte in Freiburg, inventor of the reproducing piano and manufacturer of self-playing organs and orchestrions, gave several composers the opportunity to become acquainted with self-playing instruments and to 'draw' *original compositions* on paper rolls. The punching-out of the rolls was done by experienced members of the firm. In close co-operation with Welte, the music committee of the 'Kammermusik-Aufführungen zur Förderung zeitgenössischer Tonkunst' in Donaueschingen decided to dedicate a major part of 1926-27 season (at that time the venue was temporarily Baden-Baden) to mechanical music. On 16 July 1927 one of the concerts of this chamber-music festival was even dedicated exclusively to 'original compositions for mechanical instruments', in this case mechanical piano and mechanical organ. In addition to the Fantasia in F minor for mechanical organ by W.A. Mozart, a 'Suite for Mechanical Organ' by Paul Hindemith (which has since disappeared), and a 'Study for Mechanical Organ' by Ernst Toch were performed. Also represented were Nicolai Lopatnikoff with a 'Toccata and a Scherzo for Mechanical Piano' and Hans Haass with a 6-voice Fugue, played at breakneck speed, and an Intermezzo. The first part of the 'Ballet Mécanique' by George Antheil, originally conceived for sixteen (!) self-playing pianos, was performed by one self-playing piano, since the performance of the original version for sixteen raised insuperable synchronisation problems. The same evening, Hindemith's incidental music for the film 'Sullivan, Felix, the Circus Cat' (for mechanical organ) was heard. Before the introduction of sound films, mechanical instruments had been used as accompaniment for films. Unfortunately, the mechanical organ used at the time as well as several of the original compositions for mechanical instruments have since disappeared.

With the spread of more efficient and cheaper forms of musical reproduction by around 1930 (records, radio, sound films), self-playing musical instruments sank more and more into oblivion, and they may well have stayed there had it not been for an extraordinary and genial musical hermit in Mexico who dedicated his life's work to the player piano: Conlon Nancarrow. His contributions only recently recognized by a few, Nancarrow is today regarded as one of the most significant composers of the twentieth century. In self-induced musical isolation he created a magnificent œuvre for this instrument.

Fig. 1: Conlon Nancarrow, Amsterdam 1987 (Photo: Jürgen Hocker)

NANCARROW'S PATH TO THE PLAYER PIANO

Conlon Nancarrow was born 27 October 1912 in Texarkana, Arkansas. As a child he took his first, albeit unsuccessful, piano lessons. He said once: 'this horrible piano teacher I had at the age of four ... Naturally I never learned to play anything.' Although he would later devote his life to the piano, had never actually learned to play this instrument, but did have a youthful love for jazz and learned to play the trumpet. After finishing High School in Texarkana, he went to the Vanderbilt University in Nashville to study engineering, but gave up after a few weeks. At 17 he attended the Cincinnati Conservatory, where he studied theory, composition and trumpet. But this did not satisfy him either, later commenting: '...it wasn't what I wanted, so I dropped it. I was looking for something a little less academic.'[1] He married fellow music student Helen Rigby at 19, but was divorced a few years later. At the age of twenty-one he moved to Boston, where he took private lessons with Nicolas Slonimsky, Walter Piston and Roger Sessions, who taught him counterpoint. This was the completion of his short musical training, and Nancarrow considered himself an autodidact, having acquired his musical skills through his own efforts, with the help of books and through hearing music. A decisive influence on his musical development was a performance of Stravinsky's 'Rite of Spring' which he witnessed in 1929. Then seventeen, he later recalled: '...well, it was a total revelation. At that time I'd heard practically no contemporary music, and suddenly 'The Rite of Spring' was thrown at me, and it just bowled me over. This was when I was in Cincinnati. I heard it at a concert there, and it just opened up a new world to me.'[2]

Stravinsky, Bartók and Johann Sebastian Bach were his favourite composers. Nancarrow's record collection reveals his wide-ranging musical influences, including music from Africa, India, Bali, Sumatra, China, Java, Haiti, Brasilia, Cuba, as well as an impressive collection of Jazz with Louis Armstrong, Bessie Smith and Jelly Roll Morton. There is hardly any recording of 19th century music, but music from the 17th, 18th and 20th centuries dominate the collection.

Nancarrow's first surviving compositions were written in Boston: 'Prelude' and 'Blues' for piano. Although these pieces make high demands on the performer, they can nevertheless be performed by a good pianist. At the age of 25 Nancarrow decided to join the Abraham Lincoln Brigade in Spain, fighting against the fascist Franco government. During his two years absence the two compositions 'Prelude' and 'Blues' were published in the *New Music Edition*. In 1938 the well known composer Aaron Copland wrote about Nancarrow's works in *Modern Music*:

> 'Conlon Nancarrow is a brand new name to me. I first saw it on the January 1938 cover of 'New Music', containing a 'Toccata' for violin and piano, a 'Prelude' and 'Blues' for piano alone, all by Mr. Nancarrow. His biography is brief: 'Born 1912, Texarkana, Arkansas. Studied at Cincinnati Conservatory for two years. Worked way to Europe in 1936. No job since return. Went to Spain to help fight Fascism. There is nothing to do but hope for his safe return. Otherwise America will have lost a talented composer. In fact, these short works show a remarkable surety in an unknown composer, plus a degree of invention and imagination that immediately gives him a place among our talented younger men.'[3]

After Nancarrow's return to the US he had an experience which convinced him to get rid of human performers. He had written a septet which was to be performed in New York, and as he recalled:

> 'In fact, the septet was played once in New York after I came back from Spain – I think in 1941. In any case, that was one that was played! Actually it wasn't very complicated. It had a conductor. The League of Composers had very good musicians. They got them from studios there, from the radio. There were two rehearsals. For one rehearsal, four came. The second rehearsal, three, and one of the original four. So there wasn't one session with the whole group. And when they played it, a couple of instruments lost their place right at the beginning. All through the piece, they were playing in some other place. Everything was lost, it was a real disaster.'[4]

From that time onwards, Nancarrow decided to compose for player piano (a decision reinforced by Henry Cowell's book *New Musical Resources*) in order not to be hindered by the limitations of a performer in the future. Having lost his passport in Spain, Nancarrow applied for a new one, but his application was rejected due to what the government considered his 'undesirable' activities in Spain. He then decided to emigrate to Mexico, where he lived for forty years in musical isolation, without contact with the established music scene in the United States or in Europe. About 1945 he wrote a three movement Sonatina for piano, which, however, was unplayable for a pianist because of its tremendous speed. Wanting to hear it, he arranged it for player piano. It was the first composition he ever punched in a paper roll.

Only once, in 1947, did Nancarrow leave Mexico for a few months and travelled to New York in order to acquire a punching machine. He had been looking for one for several weeks and finally became acquainted with Lawrence Cook, who arranged piano rolls for QRS. Cook lent him his hand-punching machine so that Nancarrow could copy this mechanism for punching music rolls. In his first years in Mexico Nancarrow was also busy with the idea of operating a percussion orchestra of 88 instruments by a music roll. Because of the enormous technical difficulties, this idea could not be realised.

In 1949 Nancarrow created his first original composition for player piano, which was published in 1951 in *New Music Edition*. This was to be the only score of his player piano compositions for 25 years. Nancarrow still describes his first player piano compositions as *Rhythmic Studies*, a title that he later changed to *Studies*. He composed forty-nine *Studies for Player Piano*, some of which are in several movements. Only once did Nancarrow try to organize a concert with his player pianos in Mexico, on 30 July 1962, in Sala Ponce in the Palacio de Bellas Artes in Mexico City. After he had gone through the enormous trouble of moving his two pianos to the concert hall only a few audience members turned up, most of them friends of Nancarrow, who more or less knew his music. After that Nancarrow was never willing to bring his pianos out of the studio again until October 1990, when the author and the composer Julio Estrada convinced him to agree to two player piano concerts at the University of Mexico City.

Fig. 2: Conlon Nancarrow on his punching machine, Mexico 1989 (Photo: Jörg Borchardt)

POSSIBILITIES AND LIMITATIONS OF THE PLAYER PIANO

Player pianos were brought on to the market by a number of firms. The instruments functioned, however, on the same principle: a vacuum bellow powered by a motor, sometimes also by footpedals. All other functions are pneumatically operated by means of vacuum. The piano keys are played with the help of small bellows, which are emptied of air and shut accordingly, in response to a corresponding command on the note roll. If the note is to sound louder, the bellows are quickly emptied of air. If the note is soft, the air escapes slowly.

Nancarrow used the Ampico system of the American Piano Company. The note roll is 28.5 cms wide and divided into ninety-eight tracks. These tracks are 'read off' by a tracker bar which also possesses ninety-eight holes, by means of suction. Eighty-three of these tracks control eighty-three piano notes, from the lowest 'b' to the fourth 'a' above middle 'c'. Just one track controls the left, and one track the right pedal. The keyboard is divided into two halves, which can be dynamically controlled independent of each other. On the one side of the note roll there are 6 tracks for controlling the dynamics of the bass half, on the other side the dynamics of the treble half is also controlled by 6 tracks. One track controls the rewinding action of the roll. If a perforation or a string of perforations on the note roll runs across the tracker bar, a key on the piano is struck. When the perforation is closed again, the action is ended. A single perforation means a staccato touch, and a row of perforations gives a

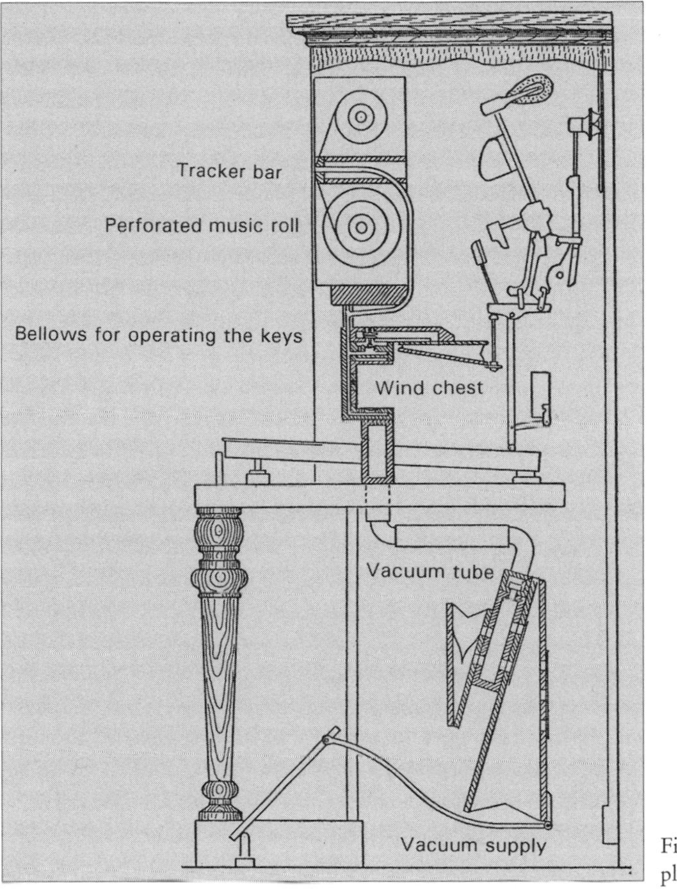

Fig. 3: Principle of a player piano

long sustained sound. The note roll can only give yes/no instructions, and in this way is an actual digital information carrier – a predecessor of the computer.

Possibilities of the Player Piano vs. the Piano Player

- *Use of the keyboard*
  A pianist can play up to 12 or 15 notes at once, which must be situated in particular parts of the keyboard. Playing the bass, treble and middle of the keyboard at the same time is not possible without assistance. The player piano can play up to 40 notes at once without the limitations of human fingers.
- *Speed*
  A virtuoso pianist can play up to fifteen notes per second in succession. With a player piano, up to two hundred notes per second are possible. This means that

one does not hear single notes, but new musical structures as sound aggregates, 'sound clouds'. Furthermore, continuous speed changes or different speeds in various lines can be reproduced.

- *Metres and rhythms*
  A pianist can play two different metres at the same time, if they are in simple time relationship with each other, such as 2 to 3 or 3 to 5. More complicated time-relationships, for example reproducing three or more different metres at a time, cannot be achieved. The same is true for extremely complex rhythms. The player piano, on the other hand, executes all complicated metres and rhythms with absolute precision. It is Nancarrow's merit that he recognised all these resources and possibilities and used them in his compositions. This led to the discovery of totally new sound structures, to the dissolution of single notes and to the emergence of 'sound layers'. Of greatest significance in Nancarrow's compositions, however, are exact time durations: temporal relationships like metre, rhythm and speed have absolute priority over melody and harmony.

'From the time I started composing, I'd always had this thing of working with temporal matters, rhythm and so forth, and this thing sort of grew. By the time I saw Cowell's book, it was just a big push ahead... I met him once. He asked me for those tapes and I sent them, and I never heard a word from him again. In fact, someone – I forgot who – pointed out that Cowell always talked about these things, polyrhythms and so forth, but neither he nor Ives ever dabbled in player pianos, which would have been the ideal way of doing that. It surprises me that he never did.'[5]

In this way, Nancarrow already used the possibilities of computer music a long time before there was a music computer. But the price he had to pay for this was high: nearly one year's work for only five minutes of music.

Nancarrow's Compositions for Player Piano

Already in the first movement of the Sonatina, written in 1945 for human player, there are frequent bar changes that can hardly be played by a pianist. A further example of the high demands made on the performer is an early work written about 1935: the Toccata for violin and piano. The piano accompaniment demands very fast note and chord repetitions. No pianist can perform these repetitions at the prescribed speed without suffering from muscle-cramp. Nancarrow therefore punched the piano accompaniment onto a note roll, and the player piano performs all these difficulties with the greatest of ease. The violin part, too, is extremely difficult, but can be played by a good violinist.

Nancarrow explained his composing process once in an interview:

'Generally, I plan ahead and write all of the Study out first, before beginning the punching. I write it down in such a way, now, that no one could really figure it out, but the first twenty Studies I wrote out in standard notation. There were rhythmic juxtapositions where measures came, but they were standard because of the fixed ratchet principle of the punching machine.

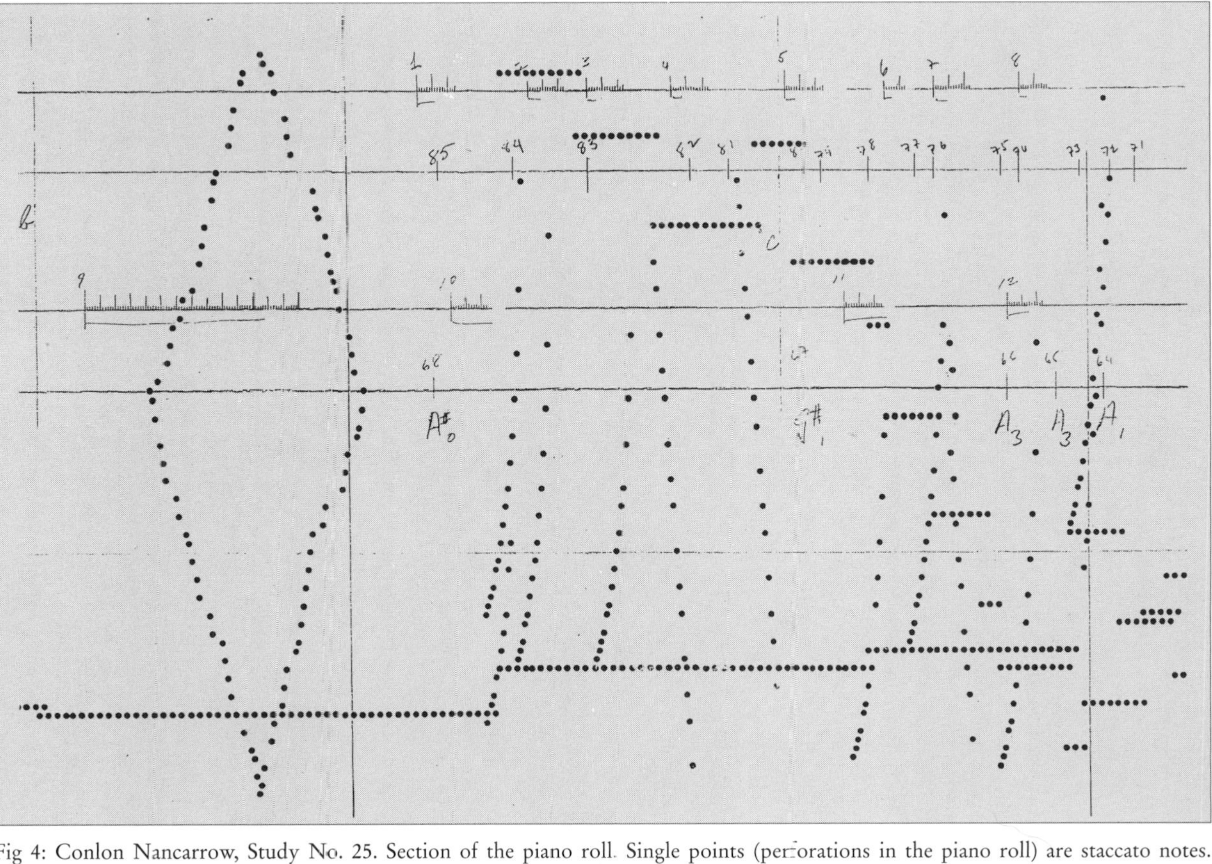

Fig 4: Conlon Nancarrow, Study No. 25. Section of the piano roll. Single points (perforations in the piano roll) are staccato notes. Horizontal rows of perforations show sustained notes. Nancarrow used different scales for the different speeds of voices.

Everything was additively related to a small unit. Now, almost all of the Studies I do are in irrational relationships. Of course, I have more or less an idea of what I am going to do and of the whole piece before I do anything: the general plan. I mark out on a blank roll of paper all of the proportional relationships of tempo, using what I think is going to be the smallest (fastest) note value as the unit of measure. Of course, occasionally, if I have to use something even faster, I just go over the roll and put in the smaller values, showing the relations to the basic scale in the score. I mark the whole thing out from the beginning to end on the blank player piano roll. It's quite accurate, I mean, as exact as I can make it. Then I take music paper and I block off the roll according to a basic size (the width of the music paper); next I take the marked proportions from the roll onto the music paper. It is not as exact as the roll, but fairly accurate so that the vertical relationship of tempo units will be more or less what I see graphically on the marked paper. I establish the pattern of temporal relationships before the pitches. The marked out roll has no rhythms, only a series of sixteen notes, or whatever. When I start to write the piece, the melody and rhythm – the harmonic connotation – are all done together. I use melody also in a rhythmic sense, in the sense of accentuating a certain rhythm by contour. But I don't sketch the melodic contours before beginning. I do the melodies and the harmonic relationship all at the same time, on the music paper that is marked out.'[6]

Nancarrow's best known player piano composition is his Study No. 21, entitled 'Canon X'. It is an example of a piece with continuous changing speed in different voices. This composition is strictly two-voiced. The bass voice begins slowly with four notes per second. Shortly after, the treble voice enters at the breakneck speed of 39 notes per second. The bass voice accelerates continuously, while the treble voice becomes slower. At about the middle of the piece the different speeds meet. The bass voice then becomes faster than the treble, and ends in a sound hurricane of 118 notes per second. The volume increases gradually during the piece in the degrees of piano to fortissimo.

Only a few of his player piano compositions can be played by hand:

'I just write a piece of music. It just happens that a lot of them are unplayable. I don't have any obsession with making things unplayable. A few of my pieces could be played quite easily – a few! In fact, Study No. 26, 'Canon- 1/1', you could play that with organ, orchestra, or any way.'[7]

One of his most impressive compositions is his Study for Player Piano No. 27. Nancarrow writes about this study:

'With No. 27 I thought of the whole piece as an ostinato, that I was going to have the exact proportions of sections worked out, before composing it. There would be a certain amount of this and a certain amount of that, but through the whole was going to be an ostinato – that against a constantly shifting acceleration and retarding. In fact, I like to think of the ostinato in that piece as the ticking of an ontological clock. The rest of it – the other lines – wandering around... There are four different percentages of acceleration and ritard that react against the ostinato: 5%, 6%, 8% and 11% ritards and accelerations. Incidentically, that's the only piece I ever did over again...'[8]

An example of different fixed speeds in various voices is Study No. 36. In this four-voiced canon all four voices are absolutely identical, except for their speeds. The first voice is running in the bass with the tempo mark 85, followed by the second with tempo 90. The third voice begins with tempo 95 and the last voice, the treble, is at tempo 100. The faster voices now follow the slowest voice, and in the middle all four voices meet together. Nancarrow's genius is evident in the way in which he combines these four voices into a great compositional unity.

> 'One reason for working with the player piano was my interest in temporally dissonant relationships. Temporal dissonance is as hard to define as tonal dissonance. I certainly would not define a temporal relation of 1 to 2 as dissonant, but I would call a 2 to 3 relation mildly dissonant, and more and more so up to the extreme of the irrational ones. When you use a canon, you are repeating the same thing melodically, so you don't have to think about it, and you can concern yourself more with temporal aspects. You simplify the melodic elements, and you can follow more the temporal material.'[9]

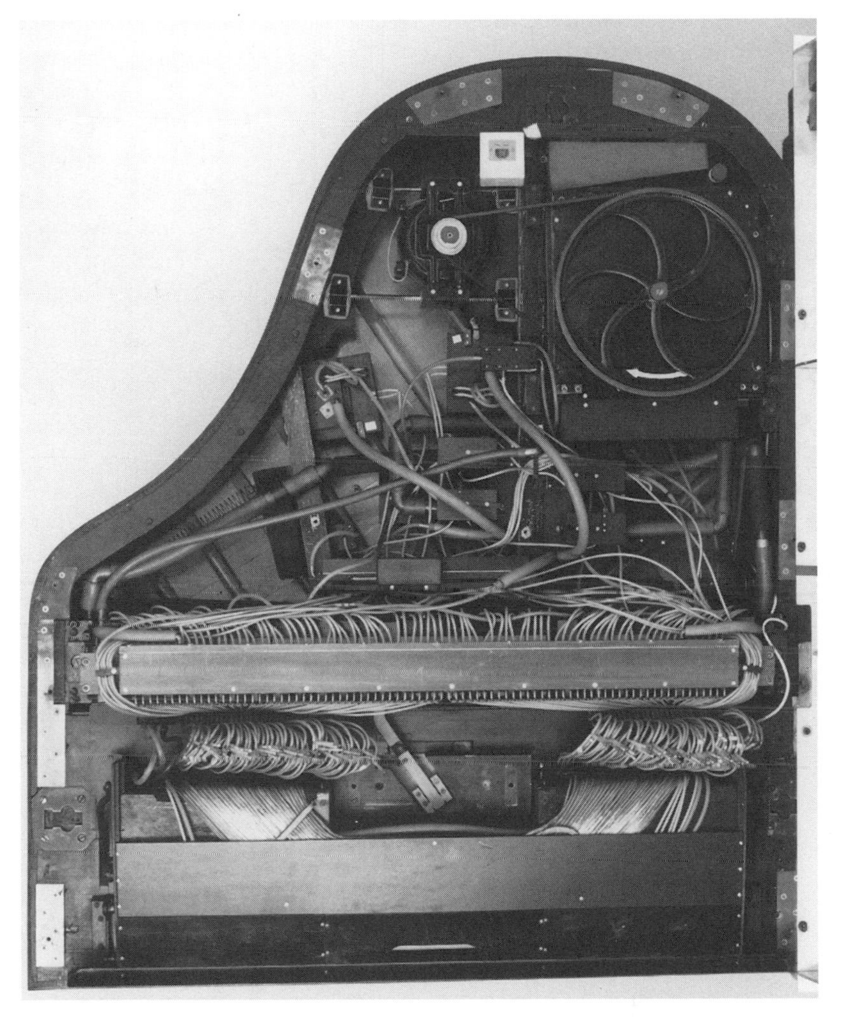

Fig. 5: The bottom of the author's Bösendorfer Grand Player Piano. Nancarrow often used this instrument for concerts in Europe (Photo: Heinrich Mehring)

For the reproduction of these highly complex compositions, Nancarrow prepared the hammers of his player pianos:

> 'In the beginning, I tried various things. The first was called a mandolin attachment. It is a wooden strip with a lot of little leather straps fixed with metallic things that dangle in front of the strings. You can lower or raise the wooden strip, and I liked the idea that you could have a normal piano or altered sound. Unfortunately, it was a mess, the leather straps were always getting tangled in the strings, especially with loud playing. Then I tried soaking the hammers in lacquer, hardening the felt. That wasn't too bad, but it wasn't what I wanted. I tried various other things; then finally settled on these: one of them has hard-wood hammers with steel straps over them and the other, felt hammers covered with leather in which are embedded the little snaps that are used in clothing. The felt cushions a little, then the leather, and then, that metallic snap.'[10]

At the end of the 1970s, Nancarrow became better known in Europe and the United States. For the first time for more than thirty years he travelled to concerts in the United States in 1981. The next year, he was invited to concerts in several places in Europe. At this time Nancarrow's Studies for Player Piano were performed by tape, because there was no suitable player piano for these complex compositions available. Since 1987, Nancarrow's player piano music has been performed in all important music centers of Europe with the author's original Ampico Bösendorfer Grand of 1927. The author had the privilege to join Nancarrow for several concerts in Europe and Mexico.

One of the most enthusiastic admirers and promoters of Nancarrow's music is the composer György Ligeti. When he first heard Nancarrow's player piano music he wrote to a friend:

> 'After the few player piano studies of Nancarrow I listened to, I affirm with all my serious judgement that Conlon Nancarrow is the absolutely greatest living composer. If J.S. Bach had grown up with blues, boogie-woogie and Latin-American music instead of the protestant choral, he would have composed like Nancarrow, i.e. Nancarrow is the synthesis of American tradition, polyphony of Bach and elegance of Stravinsky, but even much more: he is the best composer of the second half of this century.'[11]

These days Nancarrow's Studies for Player Piano are valued as the 'Well-tempered Piano of the Twentieth Century'. The composer died 10 August 1997 at the age of 84 at his home in Mexico City.

Notes

1 J. Rockwell, 'Conlon Nancarrow – Poet of the Player Piano', *The New York Times*, 28 June 1981.

2 C. Gagne and T. Caras, 'Conlon Nancarrow' in *Soundpieces: Interviews with American Composers*, (Metuchen, NJ 1982).

3 A. Copland, 'Scores and Records', *Modern Music*, Nov. 1937/June 1938, Vol. 15.

4 R. Reynolds, 'Conlon Nancarrow: Interviews in Mexico City and San Francisco', *American Music*, Summer 1984, 2/2.

5 Gagne, Caras, op. cit. (note 2).

6 Reynolds, op. cit. (note 4).

7 Gagne, Caras, op. cit. (note 2).

8 Reynolds, op. cit. (note 4).

9 Reynolds, op. cit. (note 4).

10 Reynolds, op. cit. (note 4).

11 György Ligeti to Mario Bonaventura, Letter from 28 June 1980.

Literature

1. Some publications by the author relating to Nancarrow and Mechanical Music:

– 'Ohne Grenzen – Musik für Player Piano', *Neue Zeitschrift für Musik*, 1995, 20-2.

– 'Auf der Suche nach der Präzision – Conlon Nancarrow und die Renaissance des Selbstspielklaviers', *Neue Zeitschrift für Musik*, 1986, 22-32.

– 'Mechanische Musikinstrumente' in L. Fischer, (ed.), *Die Musik in Geschichte und Gegenwart* (MGG), Vol. 2, 2. (Kassel, Stuttgart, 1996), 1710-42.

– 'Conlon Nancarrow 'Studies for Player Piano' – Il Clavicembalo ben temperato del XX secolo', *la musica, rivista di musica contemporanea*, 1985, 21: 3-9,

– 'Die Zeit als dritte Dimension- Zur Anatomie von Conlon Nancarrows 'Study No. 36 for Player Piano'', *MusikTexte, Zeitschrift für Neue Musik*, 1989, 31: 50-6.

– 'Von der Klangwolke zur Tonkaskade – Akustische Möglichkeiten des Selbstspielklaviers' in *Welt auf tönernen Füßen – Die Töne und das Hören* (Schriftenreihe Forum, Vol. 2, Kunst- und Ausstellungshalle der Bundesrepublik Deutschland, Göttingen, 1994), 401-421.

– 'Conlon Nancarrow und die Renaissance des Selbstspielklaviers', *Das Mechanische Musikinstrument, Journal der Gesellschaft für Selbstspielende Musikinstrumente*, 1985, No. 36: 10-17.

– 'My Soul is in the Machine – Conlon Nancarrows Kompositionstechnik – Die Entstehung einer Study for Player Piano', *Neue Zeitschrift für Musik*, 1998, 50-3.

– H. Moeck (ed.), *Mechanische Musikinstrumente, Fünf Jahrhunderte deutscher Musikinstrumentenbau* (Celle, 1987), 341-56.

2. H. Henck, M. Fürst-Heidtmann, 'Neues von Nancarrow' in *Neuland – Ansätze zur Musik der Gegenwart*, Vol. 2, 1981/82, 216/217; Vol. 3, 1982/83, 247-251; Vol. 5, 1984/85, 297-301.

3. M. Fürst-Heidtmann, 'Conlon Nancarrow und die Emanzipation des Tempos', *Neue Zeitschrift für Musik*, 1989, 32-38.

4. K. Gann, 'The Music of Conlon Nancarrow' in A. Whittall (ed.), *Music of the Twentieth Century* (Cambridge, 1995).

Nancarrow Studies for Player Piano are available on five CD's from Schott / Wergo:

– Conlon Nancarrow, Studies for Player Piano Vol.1 and 2.
Wergo CD 6168/2. CD 1: Studies No. 3a-e, 20, 44, 41 a,b,c.
CD 2: Studies No. 5, 6, 14, 22, 26, 31, 35, 4, 32, 37, Tango?, 40a,b.

- Conlon Nancarrow, Studies for Player Piano Vol. 3 and 4.
  Wergo CD 60166/7-50. CD 3: Studies No. 1, 2 a,b, 7, 8, 10, 15, 21, 23, 24, 25, 33, 43, 50. CD 4:Studies No.9, 11, 12, 13, 16, 17, 18, 19, 27, 28, 29, 34, 36, 46, 47.
- Conlon Nancarrow, Studies for Player Piano Vol. 5.
  Wergo CD 60165-50. CD 5: Studies No. 42, 45a-c, 48a-c, 49a-c.

Barbara Barthelmes

MUSIC AND THE CITY

The term 'city' refers to a specific kind of human settlement. With its economic basis and infrastructure it has an architecture of its own and a particular socio-cultural structure and administration.[1]

From the sociological point of view, the city has played a vital role in the development of art in general and of music in particular. At the turn of the 19th century the support for the arts by the church and aristocracy declined. With the rise of the bourgeoisie, art, music and the city entered in a new relationship in which institutions like galleries, salons, concert halls, critics or publishers played a large role. Already at a time when the city was not a theme in art, it was closely linked with it through the conditions of art production and reception. But when the city itself became a theme in art or in music, what kind of image did art or music convey?

Whenever the city appears in an aesthetic concept, the myth of the 'metropolis' is developed. This happens independent of the particular way in which it is presented, whether naturalistic or idealistic, narrative or reflective, in line with the prevailing aesthetic concepts or opposed to them. Through myth, according to Lévi-Strauss, the human being articulates intellectual contradictions and profound social conflicts. In myth man tries to solve or at least to moderate those paradoxes, which can be articulated in various codes and languages. So a myth neither depicts reality, even if it is told in a very concrete way, nor does it provide explicit advice for coping with reality. It is rather a way of reflecting that reality.

What paradoxes did the 'city myth' produce? The city as 'metropolis' is the place of apocalyptic infernos, but also of utopian visions. Just compare a painting of the German expressionist painter Ludwig Meidner with the bold constructivist visions of El Lisitzky. Often, the city is depicted as a man-devouring and inhospitable moloch, in which the individual is drowned in the masses. But at the same time, by relaxing normative values, it makes the liberation of the human being possible. The city is a focus of social injustice and alienation, but also a place of *joie de vivre* and of the pulse of life. Other paradoxical properties ascribed to the city are chaos and violence against organised power as well as the image of the dull masses versus dynamism and tempo. These contradictory images of the city have developed in the context of an elementary conflict which defines its essence: the conflict between civilisation and nature, or, to put it differently, between town and countryside. The stages of city development are the political city, the commercial and industrial city, and the complete urbanisation of Western society. This urbanisation of society is manifest in an implosion, but also in an explosion of the urban texture: there is, at the same time, urban concentration as a consequence of immigration from the countryside, but also an extension of the urban texture towards the countryside. This has resulted in a far-reaching dependence of agrarian areas on the city.[2] Starting from this, the main purpose of this article is to discuss the role of music in the development of the myth of the city.

I. Cities and music around 1900

Around 1900, several orchestral pieces were composed, the titles of which refer to particular towns: Frederick Delius' 'Paris: A Song of a Great City (Nocturne)' (1899); Edward Elgar's 'Cockaigne, Ouverture op. 40. In London Town' (1900); Charles Ives' 'Central Park in the Dark' (finished 1906), and the preludes to the second and third act 'Paris s'éveille' and 'Vers la cité lointaine' of Gustave Charpentier's opera 'Louise' (1900). These musical 'town images' were inspired by specific parts of urban topography.

In 'Paris s'éveille', the prelude to the second scene of 'Louise', Charpentier creates an acoustic impression of the awaking city at the foot of Montmartre. In the prelude to the third act 'Paris: Vers la cité lointaine' this musical description of a particular part of the city is contrasted with the depiction of the urban panorama. From a house on the Montmartre, the composer looks down at Paris, where the lights go on gradually.

In his overture 'Cockaigne: In London Town', Edward Elgar focuses on the heart of the city. 'In Paris, The Song of a Great City': Nocturne for Orchestra', Frederick Delius gives a general musical impression of Parisian nightlife, rather than experimenting with the musical form of nocturne. This is borne out by a short poem which Delius wrote and which he put at the beginning of the score.[3] Charles Ives' orchestral piece[4] not only refers to the place, Central Park, but also to the attitude – contemplation – which prevailed in the process of composition.

Contrary to paintings of a town or a landscape, which can only recreate the atmosphere of a particular moment, music is able to delineate the *course* of events. In musical compositions this aspect of development is generally marked by the times of day, sunrise or sunset, daytime or night-time.

The spatial and temporal coordinates to which the titles of the pieces allude, provide the frame in which the musical pictures develop. The means of musical portrayal are 'tone painting', which range from imitation via an associative depiction of city sounds to employing means of synaesthesia. In the music of Delius and Charpentier we find speech-like phrases imitating the Parisian *crieurs*, the street or market vendors. This is done either to characterise the Montmartre area specifically or as a characteristic of the city generally. In the case of London, the music of a solemn procession or military band, approaching slowly, passing by and disappearing again, are realistic elements suggesting the centre of London. Dances, laughter, merry music – fragments of a different reality – determine the sound picture which, in Ives' 'Central Park', is almost like in a movie, set off against the static background of strings.

All those compositions, with the exception of 'Cockaigne', develop static sounds in lower registers like a pedal tone. On the one hand the sounds reflect the never ending noise produced by the city. But on the other hand a low tone evokes the impression of darkness which corresponds with the thematic content. Linked with different degrees of intensity, especially with reduced loudness, an impression of spatial depth emerges. But the impression of closeness as well as distance can also be evoked by different means. In the case of 'Paris. Vers la Cité lointaine', for example, this is achieved by a horn which reminds the listener of hunting and forests. Transposed to the city this creates an impression of distance. Composing contrasts is another method to achieve this. In the same piece of music, passages of different moods alternate collage-like. In 'counterpointing' the deserted Central Park at night with the buoyant life in the city surrounding it, Ives, too, brings about such a contrast.

In spite of tone painting and musical realism, the purpose of this kind of music is not to present an exact image of a particular city. The city is not put into music according to its structure, but according to how it is perceived at a particular moment. The composer retains exactly those properties which constitute the city in a fleeting moment: light – the relationship between lightness and darkness – tempo and movement, and a fleeting, but continuous sound, which makes it possible to perceive the town as space. Like in impressionist paintings these compositions are atmospheric depictions of specific urban scenarios. The impression which the city leaves on the perceiving subject is translated into a musical image.

II. ART AND LIFE

Charpentier, Delius, and Elgar replace the object of contemplation – nature or a beautiful landscape – with the city. In Italian futurism, with its protagonist Luigi Russolo, the attitude of detached contemplation is given up. Wassily Kandinsky analysed this referring to the street:

> 'When being watched through window panes, street noises are reduced, movements in the street become phantoms. Through the transparent, but solid window pane, the street itself appears as detached, as something belonging to a different world. You open the door: you step out of your seclusion, get absorbed in what happens in the street, get you active in it and experience its pulsation with all your senses. The continuously changing grades of sounds and tempo reel around you, they rise, whirl-like, and suddenly collapse. The movements reel around you – a play of horizontal and vertical lines, which bend in different directions, a play of cumulating and dispersing patches of colour, which sometimes sound high, sometimes low.'[5]

This description reads as if it had been stimulated by Boccioni's painting 'Street Noise Enters the House' (1911)[6]. The city in all its dimensions, architectural-spatial, socio-economic, technical and particularly acoustical, enter man via his senses. It encompasses him completely and he succumbs to its noise and speed, longing for an extension of his faculties of perception, communication and locomotion with which nature has endowed him. Because the city is at the centre of the technical age, futurist musicians regard it as the epitome of progress. In 1914, Luigi Russolo composed 'Risveglio di una Città' ('The Awakening of the City'). In this composition he used noise instruments, 'intonarumori', to sing a hymn of praise to the city.

The attempts at reconstructing futurist noise music have, however, shown that the musical realisation of noise music falls well behind its theoretical concept. In order to explain this, I will briefly look at a text by Russolo which he wrote under the title 'I rumori della natura e della vita', published in 1916 in the collection of texts 'L'arte dei rumori'.[7] In it Russolo describes the city full of sounds and noise, or, to be more precise, the street as a microcosm of the city. More than by his music the text permits us to grasp an idea of Russolo's concept of 'city music'. He starts with a basso continuo, a pedal tone, which is produced by places, like busy roads. These are the vibrations of the street itself, not sounds caused by different vehicles. For Russolo, this basso continuo has sometimes the structure of an almost complete chord, sometimes that of an open fifth. Over this low, vibrant pedal tone, which can change according to the different streets, he uses other sounds. They are harmonic and rhythmic modulations of the pedal tone, for example the rhythmic development or horse's hoofbeats, the aggressive acceleration of car engines, the rhythmic oscillations of taxi cabs and vehicles with iron wheels. Compared to this, the almost fluent gliding and sliding of pneumatic car tyres acts as a counterpoint. To this never ending sound layer the continuous noise of the human crowd is added, an anonymous murmur that weaves a sound carpet of several distinct sounds and shouts. In the high registers the electrical wires of the trams tremble and vibrate, excited by the rattle of cars.

Apart from the aspects which the futurist Russolo and the realist Charpentier have surprisingly in common, Russolo's text also equates noise with life, and that distinguishes it from music as art: 'Each utterance of our life is accompanied by noise. Therefore noise is familiar to our ears. ... The tone, however, is not part of our life and is a thing for itself.'[8] According to this, the city is a place in which life is expressed acoustically. But it would be wrong to infer that Russolo regarded nature and the technological world of the city as mutually exclusive. For him nature is a preliminary stage to the noise of the city. He analyses nature's tone colours and rhythms (thunder, wind, rain, water, forest) in the same way in which he investigates the noise of the city. For him, the sounds of nature are limited in scope, whereas the machines in the city have the potential of bringing forth an almost infinite variety of sounds. Therefore, the idea 'nature versus technology' or 'nature versus city' is a false dichotomy: in spite of different qualities and quantities of sound, nature and city are part of life, from which art is distinct.

III. Urban Aboriginals

'Urban aboriginals'[9] is the name of a small festival which the 'Association of the Friends of Good Music' founded in Berlin in 1985. Since that year musicians from different countries have performed there regularly, such as vocalists like David Moss and Greetje Bijma, and performance artists John Rose, Harry de Wit, Stelarc, and Etant Donné.

Why 'urban aboriginals'? One of the reasons this festival came into being is the radical social and cultural change of the 1970s, which encompassed:

1. The change in political culture mainly caused by a cultural protest movement, rooted in all strata of society which came out against militarism and the unrestrained exploitation of natural resources.
2. The 'post-modern' insight that knowledge and therefore political and cultural predominance is no longer organised in universal discourses, but in a multitude of discourses.
3. The revolution in the media, the development of the personal computer and its consequences for music, the medialisation of communication and the creation of electronic networks.
4. Punk, a movement located between juvenile subculture and avant-garde art, which marked the end of a relatively homogenous sequence of styles in youth culture. Together with the music videos, it accelerated the dissolution and diversification of stylistic elements, eclecticism and the frequent use of historical quotations.

As part of these social and cultural changes, a new image of the city has come into being. In the 1970s young people responded to cities of concrete with an exodus from them, with a longing for nature and an opposition to consumerism. In the 1980s they took hold of the city; the most conspicuous movement in this context was that of the squatters. In addition to the existing primary architecture and social structures, new, secondary structures and networks emerged. William Gibson, cult author of the cyberspace-science fiction novel 'Virtual Light', describes such a secondary structure.

> 'The integrity of its span (the bridge's span) was rigorous as the modern program itself, yet around this had grown another reality, intent upon its own agenda. This had occurred piecemeal, to no set plan, employing every imaginable technique and material. The result was something amorphous, startlingly organic ... Its steel bones, its stranded tendons, were lost within an accretion of dreams: tattoo parlours, gaming arcades, dimly lit stalls stacked with decaying magazines, sellers of fireworks, of cut bait, betting shops, sushi-bars, unlicensed pawnbrokers, herbalists, barbers, bars. Dreams of commerce, their locations generally corresponding with the decks that had once carried vehicular traffic; while above them rising to the very peaks of the cable tower, lifted the intricately suspended barrio, with its unnumbered population and its zones of more private fantasy.' [10]

This image of a new urban structure correlates with the conception of music held by a generation whose musical socialization took place under the influence of the medialisation of music and a multitude of different musical cultures. They therefore broke away from the traditional hierarchies of music, such as Roberto Paci Daló and Joel Rubin[11] have done, who combined the results of ethnological research on

Heroner and hassidic melodies from Israel with an interactive live computer system. This generation also no longer makes a distinction between 'higher' and 'lower', between 'popular' and 'classical'. In his piece 'Forbidden Fruit' John Zorn, for example, combines a string quartet with turntable and voice. To produce those sounds, artists often use materials from the remnants of our consumer society, from garbage, the scrap yard or from the reservoir of used, musical apparatuses. Fast Forward uses Caribbean steel drums, wind instruments made out of plastic tubes, broken glass and pneumatic hammers. For his 'electrophonic orchestra' Gordin Monahan applies various metal parts of different origin. Those artists should also be mentioned who, more than 40 years after John Cage's 'Imaginary Landscape No. 1', used the record player as a musical instrument.

These artists have a similar attitude towards new music technologies. They favour hybrid configurations combining traditional instruments with high-tech electronics. Nick Collins' 'trombone-propelled-electronic' is a prime example. A microprocessor from a Commodore 64 is, together with a hypersensitive echo set, combined with an old trombone. But in spite of all the electronics which are generally used 'live', Collins and others regard music making as a tactile, physical event, in which the direct relationship between physical effort and sound production is restored. Therefore these sound artists do not create musical works in the traditional sense, but rather musical forms, which break up and visualise the traditional character of a musical work. Often their own bodies play a vital role in the performance.

The term 'urban aboriginals' is taken from ethnological thinking and was transferred to the aesthetic practice of our time. On the one hand it implies the view that contemporary music is hidden in the city jungle and that one should investigate the culture and art of the city as one would research the customs of aboriginal tribes in the Australian bush. On the other hand the term alludes, in an ironical way, to the seminal research of the French ethnologist Claude Lévi-Strauss on 'The Savage Mind'.[12] In his description of the *bricoleur*,[13] he gave *bricolage* the same status as traditional scientific practice and was thus wholly in line with postmodern thinking. The futurists view the city as nature created by man. The city of today is characterized by a structure which breaks with modernism and has produced a new type of artist – the urban aboriginal.

IV. Soundscape

Today the term 'soundscape' is often used in the context of sound installations and acoustic environment. Apart from this rather unspecific definition, there are others which bring it closer to the ecological, sociological, and semantic conceptions of sound.

Justin Winkler, for example, distinguishes the term soundscape from connotations which refer to the landscape and its 'green' aspects. In this context the concept of soundscape is not restricted to a musical representation of 'outer nature', as contrasted to mechanically, electrically or electrophonically produced sounds. Instead of looking at isolated 'sound biotopes', it denotes the totality of environmental sound aspects from the perspective of perception. This environment extends from the noise ones own body makes to sounds faintly audible in the distance.[14] The term soundscape also makes clear that we are dealing with physical sounds. They have become part of our environment, because they are perceived by us.

At the end of the 1960s, R. Murray Schafer and Michael Southwort independently started to use the term soundscape in reference to the city as an urban complex. With Schafer the term is part of his concept of acoustic ecology. This aims at a critique of the acoustical appearance of our predominantly urban environment, which he investigates in regard to its ecological status. According to Schafer, the former abundance of different sounds has become levelled out and coarsened by urbanisation, industrialisation and the destruction of nature. The noises of cars, construction machinery and aeroplanes have spread like an acoustic haze over the acoustic environment. From the acoustic-ecological point of view this has resulted in a far-reaching levelling out of soundscapes.

Apart from the concept of acoustic ecology which is still in the tradition of the hippie movement with its longing for nature and an exodus from the city, I want to introduce a different, an acoustic-sociological definition of soundscape. By analysing its acoustical aspects, its tonescapes and soundscapes, the two authors Hans-Peter Meier-Dallach and Hanna Meier treat the city as a social organism. They start from the hypothesis that sounds and the urban soundscape function as signs and refer to a specific relationship between people and the urban space in which they live. 'The sound of a city contains images of the social space and its rhythms, according to which society moves. ... It is possible to analyse the culture and values of society by examining its sounds.'[15]

Artists, who create sound installations or acoustic environments in an urban, public space, refer to this social and cultural organism mediated by sound. Places, where the exchange of goods and human mobility is particularly intensive and fast, are not only represented by a specific architectural form, but also by specific auditive signs. As icons of our mobile global society, train stations and airports are such places. When reconstructing the sound of the former Anhalter railway station in Berlin in 1984, Bill Fontana referred to the urban sound context of a railway station. Eight loudspeakers emitted sounds which had been recorded earlier at three different points of the Cologne railway station.

The soundscape of a modern major airport is completely different from this, arising from access roads, entrances, corridors, information and refreshment areas. On a sound carpet consisting of hardly audible runways, automatic doors and monitors, the loudspeakers announcing the arriving and departing flights can be heard only mutedly. 'Ambient 1: Music for Airports' (1978) by Brian Eno adapts itself well to this low sound level. With its slowly evolving sound carpets and repetitive patterns it tries to introduce islands of quietness into the course of events from entering the building until the departure of the plane. In 1996, as part of the exhibition

'Sonambiente' which was part of the 300 anniversary celebrations of the foundation of the Berlin Academy of Arts, the Japanese sound artist Akio Suzuki marked points of hearing at the Spree island in the middle of Berlin, which is dominated by building sites. He suggested that visitors should stop at these places and open their senses to the sound of the city. In the minds of those who did this, a kind of acoustic city map of the Berlin Museum Island and the adjacent areas developed. So Suzuki supplemented the architectural identity of the city with an acoustical dimension.

At the same exhibition, Christina Kubisch's installation 'On Silence' could be seen and heard. It was located in the former wine tavern Hut, in the architectural remainder of the former Leipzig Square which has today disappeared within the largest building site in Europe. Two microphones installed at the outer windows of the room were directed towards Potsdam Square and transmitted the noise of the building site to a room inside the building. A timer divided the noise into regular intervals. The room itself was darkened and fragments of texts could be seen, dealing with the theme of silence. Because of the strong contrast between the moments of peace and the breathtaking noise transmitted to the room, that noise became the acoustic symbol of the profound changes in the city and the painful and frightening cuts into its existing texture.

If we agree that myth is an image in which contradictory experiences are digested, what images are then depicted by the different kinds of music dealt with above? One of the general patterns of the musically articulated city shows it as nature, either as a beautiful landscape created by an aestheticised view of the city or, contrary to this, as an uncontrollable city jungle. The city as a living environment is marked by soundscapes as well as by masses of houses. It is also a sign of the myth of the city that it is regarded as the technical extension of the human being. It is a giant tool, a prosthesis replacing the senses of perception and organs of locomotion through which man creates his world. Early in this century the idea prevailed that new machines would continue to produce a world of sounds and noises *ad infinitum*. Man listened to the sound of a machine as if it was birdsong. At the end of this century, however, the city has become a large stock of remainders, which the former champions of a reformation of music through the machine have recycled in an attempt to turn old into new. The continuation of the city myth is also inscribed in the human condition of life which is expressed by the word 'home'. The reason why several town districts like Soho or Greenwich Village in New York, Kreuzberg or Prenzlauer Berg in Berlin, are endowed with a particular, homely atmosphere is that their denizens have created new ways of living which suited them and which were to an extent independent of the prevailing architectural and social conditions. Music has played a large role in this adaptation. To make the city a home for its inhabitants it needs symbols which generate a common identity. Examples are characteristic places like Central Park in New York or the Anhalter Railway Station and the Museum Island in Berlin. Music enhances the mythical character of these places.

Notes

1 This text is also published in an altered and German version in: R. Kopiez, B. Barthelmes et al. (ed.), *Musikwissenschaft zwischen Kunst, Ästhetik und Experiment. Festschrift Helga de la Motte-Haber zum 60. Geburtstag*. (Würzburg, 1998), 29-40.

2 H. Lefèbvre, *Die Revolution der Städte* (Frankfurt/M., 1976).

3 Mysterious city-/City of pleasures/Of gay music and dancing/Of painted and beautiful women-/Wondrous city,/Unveiling but to those who/shunning day/Live through the night/ And return home/To the sound of awakening streets/And the rising dawn.

4 According to Kirkpatrick's register the full title is 'A Contemplation of nothing serious or Central Park in the Dark in the good old Summer Time.'

5 W. Kandinsky, *Punkt und Linie zur Fläche* (Bern-Bümpliz, 1973), 13.

6 This painting is in the Sprengel Museum Hannover.

7 This collection of texts is in: G.F. Maffina, *Luigi Russolo e l'Arte dei rumori* (Turin, 1978).

8 Luigi Russolo, 'l'Arte dei rumori' in H. Schmidt-Bergmann, *Futurismus. Geschichte, Ästhetik, Dokumente* (Hamburg, 1993), 239.

9 Urban aboriginals is an ironical derivation from the English adjective 'aboriginal'.

10 W. Gibson, *Virtual Light* (London, 1993), 58 f.

11 Roberto Paci Daló is a composer and clarinet player; Joel Rubin is a musicologist and klezmer clarinet virtuoso.

12 First published in Paris in 1962.

13 In the English edition of C. Lévi-Strauss, *The Savage Mind* (Oxford, New York, 1972), 17, the translator notes: "The 'bricoleur' has no precise equivalent in English. He is a man who undertakes odd jobs and is a Jack of all Trades or a kind of professional do-it-your-self man, but, as the text makes clear, he is of a different standing from, for instance, the English 'odd job man' or handyman."

14 J. Winkler, 'Beobachtungen zu den Horizonten der Klanglandschaft' in G. Böhme, G. Schiemann (eds.), *Phänomenologie der Natur* (Frankfurt/M., 1997), 274.

15 H.-P. Meier-Dallach, H. Meier, 'Die Stadt als Tonlandschaft' in *Gesellschaft und Musik. Wege zur Musiksoziologie. Festgabe für Robert H. Reichhardt zum 65. Geburtstag* (Berlin, 1992), 416.

HANS-JOACHIM BRAUN

"MOVIN' ON": TRAINS AND PLANES AS A THEME IN MUSIC

I

> On October 4, 1923, I played in Paris for the first time, almost exactly a year behind my first scheduled appearance for Paris. My little group of piano pieces, the 'Mechanisms', the 'Airplane Sonata', and the 'Sonata Sauvage' were to go on as a prelude to the opening of the brilliant Ballet Suédois (Swedish Ballet) which Rolf de Mare was bringing to Paris this season for the first time. .... My piano was wheeled out on the front of the stage, before the huge Leger cubist curtain, and I commenced playing. Rioting broke out almost immediately. I remember Man Ray punching somebody in the nose in the front row. Marcel Duchamp was arguing loudly with somebody in the second row. In a box nearby Erik Satie was shouting, 'What Precision! What Precision!' and applauding.[1]

This is an account George Antheil, avant-garde composer, piano virtuoso, and self-styled 'bad boy of music', gave of one of his concerts in Paris in the early 1920s. The somewhat excited response was not unusual. A few years before, the Italian Futurists and Igor Stravinsky, among others, had met with similar reactions.[2] A major theme in the work of Antheil's or the Futurists was the machine, particularly transport technology like cars, railway engines or aeroplanes. Theirs was an age of speed, of power, of attempts to break the barriers of time and space. Their often provocative performances disturbed more conservative listeners. Many adherents of new music turned concerts into dada-like happenings and were never adverse to a good fight. Brawls ensued, with the details faithfully reported in the press. But the musical representation of railway engines and aeroplanes did not always generate such controversy. Composers like Arthur Honegger managed to make those means of transport more palatable to the general audience. In their music, which is still popular today, they convey the impression of energy, drive and elegance associated with railway engines and aeroplanes.

The purpose of this article is to trace the origins of transport technology as a theme in music and to explore the contents of and responses to some of the more prominent pieces. In doing so I will consider whether there was a prevalent view of technology and if so, whether it changed over time or remained comparatively stable.

II

If in the 18th and 19th centuries composers made explicit references to their environment they generally meant nature and the way nature affected human life. Beethoven's 6th Symphony, the 'Pastorale', is a well-known example.[3] But there are exceptions like Hector Berlioz's 'Song of the Railway' which he wrote in 1846.[4] With the development of industrialization an increasing number of composers no longer considered it as appropriate to obtain their thematic inspirations from a pre-industrial era. They now took production processes in factories and new means of transportation like automobiles and aeroplanes into consideration.[5]

The emergence of railways in the first half of the 19th century met with different responses. There was a lot of enthusiasm and the feeling that railways extended the possibilities of human life. Increased mobility meant economic growth and social opportunities. But there were also mixed feelings and pessimism. Not everybody welcomed the prospect of a life dominated by factories; many were suspicious of a transport system geared to transform a pastoral landscape into one dominated by industry. High expectations as well as trepidations and apprehensions are reflected in 19th century literature and fine arts.[6] Although one should be wary of making simplistic connections between the pace of life and the tempo of music, many composers of the early 20th century consciously wrote music to reflect a changing world.[7]

In the 1920s railways and particularly locomotives aroused the fascination of many composers. This is particularly true of the Franco-Swiss composer Arthur Honegger, who in 1923 named his symphonic movement 'Pacific 231' after one of the fastest American locomotives of its time. In 'Pacific 231' he successfully transformed features like speed, dynamics and energy into the language of music. The music starts very slowly and moves gradually up to full speed. Then the process reverses. Interwoven into all this is a mighty chorale in the Bach manner. Although the composer repeatedly talked about his love of locomotives he maintained that 'Pacific 231' was no 'program music' in the strict sense and insisted that it was instead the translation of a visual impression and of physical joy into music.[8]

Given their insistence on rhythm it is not surprising that jazz and blues musicians found inspiration in locomotives and railways. The blues, for example, has immortalized the 'Cannonball', the Illinois Central Express from Chicago to New Orleans, and title train in Bessie Smith's 'Dixie Flyer Blues'.[9]

But energy and drive are not really typical of the blues genre. For this one has to look at jazz, especially at some big band pieces of the Swing era. In this context Duke Ellington's 'Daybreak Express' (1933) stands out and can perhaps be called the greatest train-inspired piece in music. Similar to 'Pacific 231' the train gradually accelerates, picks up steam and then barrels down the track at breakneck speed. At the end of the journey it slows down and stops with a dissonant wheeze.[10] In post-war pop music the German group 'Kraftwerk' with a piece like 'Trans Europe Express' (1977) reflected the same fascination with the theme of transport.[11]

Closely connected with energy and power is the idea of liberation and freedom through railways, aeroplanes and automobiles, but also through technology in general. A fitting example of a jazz piece is John Coltrane's 'Song of the Underground Railroad' (1961) with the double meaning of underground transport and an escape route from slavery.[12] Getting away to a better place, sometimes to heaven on the

Arthur Honegger 1892-1992, Katalog der Ausstellung, Le Havre, Zürich 1992, ed. Stadt Le Havre and Schweizer Musikrat (Zürich, 1992), 54.

gospel train, getting on board to ride to freedom and to the promised land, sometimes with the 'engineer Jesus' as guidance, are also familiar motifs in blues.[13]
But technology means different things to different social groups and to different individuals. In the case of the railway as a blues theme there is often freedom, hope and excitement for the male and frustration and hopelessness for the female. The male – my man, my daddy – goes away to the North to get a job, his 'mama' stays at home, often with a sizeable number of children, and will probably never see him again.

> 'I'm going up North, baby, I can't carry you.
> Ain't nothing in that cold country a sweet girl can do.'[14]

However, this gender-related point should not be laboured too much. As Angela Davis has pointed out, African-American women could also associate themselves with travel as a mode of freedom. Their role was not only passive. The (female) protagonist of 'Traveling Blues', for example, does not know where she is headed, but she purchases a train ticket and thus refutes the blues cliché that

> 'when a man get the blues he hops on the train and rides,
> but when a woman get the blues, she lays down and cries.'[15]

Swing pieces are of a different nature. Billy Strayhorn's 'Take the A Train' (1941) with its unforgettable and inspired theme featuring a catching augmented chord in the third and fourth bars became a leitmotif of the swing era. Ellington adopted it as his orchestra's signature tune from 1941 onwards.[16]

In the examples mentioned above, technology is often heroic and larger – than – life; the railway engines are impressive and inspiring technological giants that liberate people by providing freedom of movement. There is, however, another characterization of the railway in popular song, one which emphasizes a friendlier, cosier relationship between riders and machine. In these cases technology is not 'the other', railway engines are not qualitatively different from living things. They are no longer giants of iron and steel, but friendly helpers of mankind, a cross breed between beasts of burden and pets. This is true of the Brazilian composer Heitor Villa-Lobos's 'O Tremzinko da Caipira' ('The Little Train of Caipira') written in 1923 which is the fourth movement of his 'Bachianas Brasileras' No. 2.[17] Villa-Lobos allegedly wrote this movement in one hour while riding on a country railway that carried berry pickers and farm workers from São Paulo into the Brazilian interior. The music reflects the chugging, clanking, whistle-blowing train journey that inspired the composer's imagination.[18] Popular tunes in a similar vein are Meade 'Lux' Lewis' 'Honky Tonk Train Blues', Harry Warren's 'Chattanooga Choo Choo', immortalized by Glenn Miller, or Ellington's 'Happy Go Lucky Local' (1946) part of his 'Deep South Suite' and a homely counterpoint to his mighty 'Daybreak Express'. This friendly local is quite a different train from the 'Express', as it bounces, toots, chugs, clanks and reels, bearing some resemblance to Villa-Lobos's 'Little Train of Caipira'.[19]

In many of the pieces just mentioned musical instruments are used to imitate the sound of railway engines. It is therefore not surprising that composers of *musique concrète*, based on transformed recorded everyday sound, found ample material in the different means of transport.[20] Pierre Schaeffer's first work, 'Etudes aux chemins de fer', was constructed from recordings made at the depot for the Gare des Batignolles in Paris and includes the sound of six steam locomotives, of their whistling, of trains accelerating and of wagons passing over joints in the rails.[21] The piece starts with the acceleration of a train and the sound of wagons which are put together from different recordings. By using ingenious cuts and montage Schaeffer transforms the recorded ostinato-like railway sound into some rhythmically organized sound which comes close to music in the conventional sense of the word. There is even an element of 'swing' in this. But most important to Schaeffer was the ambivalence between realistic ('concrete') sound on the one hand and music on the other, an ambivalence which other composers influenced by Schaeffer also dealt with.

In 1970 Bernard Parmegiani, a member of the experimental studio founded by Schaeffer and his group GRM (Groupe de Recherches Musicales), wrote his piece

'L'œil écoute'. It begins seemingly 'realistically' with a train at full speed, but then that train – surrealistically and within a spur of a moment – comes to a sudden halt. After a whistle, it is suddenly at full speed again. As the piece continues it comes closer and closer to the idea of conventional music and even finishes with a tonal chord; musique concrète gradually becomes more abstract. Parmegianis's objective is to make the listener think about the ambivalence of everyday noise, sound and music as organized sound.[22]

The composer Steve Reich also uses concrete train noises. 'Different Trains' (1988) is about a train between New York and Los Angeles which he rode in his childhood during the Second World War, while at the same time as a Jew in Europe he would have had to ride a different train, possibly leading him to his death in a gas chamber.[23]

The notion of power, energy and drive has little room for failure. But what happens when something does go wrong? In listening to Gioacchino Rossini's 'Little Holiday Train' (1864) the first impression is of a pleasant, leisurely train ride. But then the train derails and causes casualties. The idyllic train ride thus comes to a sudden, terrible end.[24] In popular music, in jazz, blues and especially in folksongs, there are numerous cases of train accidents. Scott Joplin's 'Crush Collision March' (1896) is an early example of programmatic musical with its dissonant frantic train whistle and 'daring' seventh chord during the collision.[25]

In most of those cases the accidents described are due to mechanical failures, to human oversight and negligence, but also to daring and even foolhardiness. In the ballad of the 'Altoona Freight Wreck' (29 November 1925) on the Pennsylvania Railroad the engineer started the train without checking that the air-brake system was working properly and he never regained control.[26] On the 'Wreck of the C&O Number Five' the train hit a broken rail. A misunderstanding between the road foreman of engines and engineer caused the 'Wreck of the Royal Palm' and the conductor's misreading of the meet order led to the 'New Market Wreck'. A very common cause of accidents was the attempt to make up for lost time. This sent 'Casey Jones, the Brave Engineer', an American folk hero, known as a 'fast roller', to his early grave, taking many passengers with him. Something similar caused the 'Wreck of the C&O':

> 'It's over the road I mean to fly with speed unknown to all
> and when I blow for the stock yard gates, they'll surely hear my call.'

III

At the turn of the century artists perceived the invention of the aeroplane as an aesthetic event with far-reading implications for future artistic and moral sensibility. Even more than with railways, artists and musicians, through the play of their imagination, transformed the technological event of inventing and developing the flying machine into a form of spiritual creation.[27] Similar to railways of the 19th century, but even to a higher degree, aeroplanes were supposed to bring about almost unlimited individual mobility. Americans viewed mechanical flight as portending a new, wonderful era of peace, prosperity and harmony. It heralded an age in which the promise of the 'winged gospel', a second ascension of Christ, would come true.[28] The aeroplane opened up a new dimension in human activity. Aeroplanes transformed man's sense of time and space, transcended geography, and knitted together nations and peoples.[29] According to this 'internationalist gospel', most common before 1914, war was too terrible to contemplate. Writers and artists depicted the airman as a romantic and chivalric figure, an image which remained valid in the decades to come.[30] But already before the First World War it became clear that the aeroplane could also be put to other uses: the utopia of peaceful internationalism gave way to aggressive nationalism with the aeroplane as a devastating, lethal weapon.[31]

Tommaso Marinetti, the Italian Futurist writer, saw these different options clearly. He hinted to the aeroplane's uses as a means of liberation from the past, but also as a bomb-dropping engine of war. It was Marinetti who gave Francesco Balilla Pratella, the Futurist composer, financial support while the latter was working on his 'L'aviatore Dro' (1911-14). Marinetti insisted on Pratella's incorporating 'intonarumori', noise machines, into his work. Having no proper concept of integrating noise into his composition, Pratella complied only reluctantly.[32] Although Pratella is known as a prominent Futurist composer writing Futurist manifestos, he was only half-hearted in his futurism, gradually withdrawing from the movement after the First World War. When his reputation as a conductor forced him to travel as far afield as Paris, he retained a very un-Futurist aversion to travel and speed. In his 'L'aviatore Dro' the aeroplane serves as a symbol of spiritual ascension.[33] The music relies on constant repetition of short phrases and on the whole-tone scale. Pratella uses the noise intoners to real effect when the hero's plane crashes to the ground.[34]

Heroism, speed, destruction and death were ever-recurring themes of Futurist writers, painters and composers shortly before and during the First World War. Leo Ornstein, an American composer and outstanding concert pianist with a Jewish-Russian background, felt close to Futurist ideas which he sometimes mixed with a strong dose of dadaism. Like many of his other compositions, his short piano piece 'Suicide in an Airplane' (1913) is programmatic and expressive with short percussive phrases repeated at high speed in order to achieve a feeling of motion and acceleration. Ornstein uses a short bass figure throughout the work to be played very fast, simulating the sound of an aero engine and the feeling of movement. This technique bears a close resemblance to the repeated lines of the bent leg in Marcel Duchamp's painting 'Nude descending a Staircase' which achieves the illusion of motion by repetition.[35]

In popular songs on aeroplanes the motif of escapism stands out: to get away from the problems down below, from unpleasant obligations like paying taxes or having to cope with an ill-liked mother-in-law.[36]

A page in Leo Ornstein's hand from the score of *Suicide in an Airplane* (1913). Vivian Perlis, 'The Futurist Music of Leo Ornstein', *Notes of the Quarterly Journal of the Music Library Association*, 1975, 31, No. 4: 742.

To George Antheil, the airoplane seemed to be most indicative of the kind of future to which he wanted to escape. His 'Airplane Sonata' (1921) reflects the prospect of dreams in which he seemed to have caught the true significance and atmosphere of those giant engines that move mankind about. Antheil felt that for most people the First World War had killed illusion and sentimentality. In his 'Airplane Sonata' the lack of dynamics and articulation indicators makes the music sound tough-as-steel; the first movement designation 'to be played as fast as possible' gives the piece a frenetic quality.[37]

In his 'Airplane Sonata' Antheil's machine aesthetic manifests itself in driving rhythms and insistent ostinatos. The composer constructed his sonata out of the addition and manipulation of rhythmically activated musical blocks delineated by different ostinato patterns. These musical blocks, 'time space components' which he superimposes on his musical canvas, derive their energy from the rhythmic momentum of repeated musical fragments. Throughout the sonata Antheil blocks the multimetric and polychordal fragments like a cubist painter arranges geometrical abstractions on a canvas.[38]

During the First World War technology showed its destructive potential, but after the war technological optimism soon gained ground. Motorization, majestic and fast ships (despite the Titanic disaster of 1912) and aviation caught the imagination of contemporary composers. It is therefore not surprising that the aviators of the time – and this includes 'fighter aces' of the First World War – enjoyed an almost mythical veneration. Among the pilots of the 1920s Charles Lindbergh stands out. Popular songs celebrating his first solo Atlantic Ocean flight of 1927 abound and even a dance, the acrobatic and rhythmic 'Lindy Hop', was named after him. In the late 1920s Harlem's Savoy Ballroom became an attraction to white Americans who went there to watch their African-American countrymen dancing the 'Lindy Hop'.[39]

Many of the songs on Lindbergh were written in a patriotic vein, verging on nationalistic sentiments. Howard Johnson and Al Sherman hailed 'Lindbergh: The Eagle of the U.S.A.' and James Miranto sang 'America, Lindbergh is your boy.' But other songs stressed the example Lindbergh gave to the whole world and the concern, citizens of many different nation expressed for the well-being of the daring pilot.

In his 'Lindberghflug' (1929) the German writer Bertold Brecht emphasized Lindbergh's struggle against the elements and his own fatigue. According to Brecht, his victory had to be understood as a collective triumph of the human spirit. Brecht's 'Lindbergh Flight' is a fitting example of the enthusiasm of the 1920s for the heroic extension of the human spirit through new technologies.[40] By transforming Lindbergh's flight into a didactic ballad for chorus and individual voices, with music by Kurt Weill (for the American scenes) and Paul Hindemith (for the ocean and European scenes), Brecht intended to depict Lindbergh as an example of courage which everybody should emulate. Although Brecht's 'Lindbergh Flight' is a radio cantata, it has a more ambitious aim, assigning a novel role to the medium of radio: he designed it for audience participation and conceived the 'Lindbergh Flight' as a duet between a group of participating listeners and the radio which was to broadcast the music, sound effects and part of the chorus. Although Brecht's 'Lindbergh Flight' was quite popular at the time – and even enjoys a renewed popularity today – its author was not very happy with it. He realized that the new genre of radio play with listener participation and the discussion of social issues could only be carried through within a new social system which was no longer a 'distribution system for bourgeois society' but a true democratic communication system, in which everybody could participate with equal weight.[41] He also changed the title from 'Lindbergh Flight' to 'Flight of the Lindberghs' and then to 'The Ocean Flight', omitting Lindbergh's name, because the pilot had been sympathetic towards National Socialism.[42]

IV

The theme of flight also intrigued Arthur Honegger, who had already shown his fondness for technological themes in his 'Pacific 231.' In 1943 he wrote the music to Louis Cuny's film 'Mermoz', dealing with the pilot Jean Mermoz, who in 1930 was the first to cross the South Atlantic by plane. In Honegger's score traces of the 'Lindbergh Flight' by Weill and Hindemith, as well as Antoine de Saint-Exupéry's 'Voil de nuit' are evident. The pilot is depicted as an individual with courage, but also with fears.[43] Arthur Honegger, Paul Hindemith, Kurt Weill and others also inspired Bohuslav Martinu, the Czechoslovak composer, to write his symphonic scherzo 'Thunderbolt P-47' at the end of the Second World War. It had been commissioned by the National Symphony Orchestra in Washington and was dedicated to the 'Republic P-47' 'Thunderbolt' fighter and later fighter-bomber, one of the most accomplished aeroplanes of the Second World War. In 1941 Martinu had to leave Europe as a refugee and settled in New York. His 'Thunderbolt P-47', a dramatic and partly spectacular piece of programme music, is an homage to the victorious U.S. air force.[44]

Between 1942 and 1945 several American composers made patriotic contributions to the war effort. These include Samuel Barber's Second Symphony, dedicated to the U.S. Army Air Corps, Elie Siegmeister's 'Freedom Train', and also Marc Blitzstein's Symphony 'The Airborne.' In 1943 the U.S. Army Air Force commissioned the composer Samuel Barber to write a symphony to boost morale. For several weeks the composer was flown from one airfield to another to absorbe the feeling and atmosphere of his subject.[45]

The symphony was supposed to be modern. The Colonel at West Point, to whom Barber had to report progress, stated that since the Army Air Force employed all the latest technical developments in flying, a symphony written for it should be equally up-to-date technically and musically. The composer accepted the suggestion to use an electronic tone-generator built by the Bell Telephone Laboratories to simulate the sound of a radio beam guiding the navigator on his way. This tone-generator consisted of 100 acoustic discs side-by-side and rotating together. Its signal was picked up at the rims and fed through amplifyers and alternators to loudspeakers. But the device gave great problems and Barber replaced it with an E-flat-clarinet.[46]

Barber's Second Symphony exhibits an emotional climate of great tension and energy. This is accomplished by persistent ostinati, dotted rhythms, dissonant intervals and angular lines. Throbbing ostinati and the whirring of strings evoke an atmosphere of preparing for battle. The calm of the andante can be interpreted as expressing the beauty and solitude of night flight.[47]

In this symphony which was premiered by Serge Koussevitsky and the Boston Symphony Orchestra on 3 March 1944, Barber wanted to express the sensation of flying. But he was concerned that the work should not be considered programme music in the conventional sense. Unsatisfied with it he withdraw the score from circulation in 1964. Barber found its musical quality wanting and was of the opinion that times of cataclysm are rarely conducive to the creation of good music. But he salvaged the second movement of the symphony as a seven-and-a half-minute tone poem, 'Night Flight', op. 19 a, thinking that the lyrical voice, expressing the dilemma of the individual, might still be of relevance.[48]

Barber's fellow composer Marc Blitzstein composed his Symphony 'The Airborne' during the years 1943-4, but the score was lost and he had to rewrite most of it. When the score turned up again he preferred the new one; the premiere was on 1 April 1946 under Leonard Bernstein.[49] Blitzstein's Symphony is an epic of flight. The first movement portrays man's many failed attempts to fly and ends with a salute to the Wright brothers. The second movement demonstrates that flight and aeroplanes, like many other artefacts of technology, also have a destructive potential. This becomes evident in the 'Ballad of the Cities', when Blitzstein evokes the experience of seven large cities bombed during the Second World War. The final section depicts the arrival of the Americans in Europe.

The third movement includes 'Night music. Ballad of the bombardiers' with a solo baritone singing about a pale nineteen-year-old bombardier writing a letter home to his girlfriend. The actual 'job' of the bombardiers is not the theme here, but his loneliness when the job is done. Generally, it was much more difficult for the lyricists to present the crew of a bomber in an attractive way than the fighter crew. With the fighter crew there was daring, bravery and skill, in the bomber the efficient cooperation between the crew and the machine, verging on a man-machine symbiosis, was highlighted:

> 'A guy twenty four and a B-twenty nine
> like two sweethearts, they're always a pair
> What a team to take care – of the air –
> Anywhere – Every – where.'[50]

Blitzstein's 'Airborne Symphony' is a theatrical piece. Something similar can also be said of Karlheinz Stockhausen's 'Helicopter Quartett' (1992-3), a piece by a much more demanding composer and one of the main protagonists of contemporary music, whose seminal 'avantgarde' ideas have had an enormous effect on 'art music' as well as on popular music. The 'Helicopter Quartett for string quartet and helicopters' is part of the composer's gigantic cycle 'Mittwoch aus Licht'. The four musicians board a helicopter and play as they are ascending and descending. The music of the string quartet mixes with the noise of the helicopter engines and is broadcast to the concert hall.[51] The 'Helicopter Quartet', not unlike some compositions of 'musique concrète' which Stockhausen did not regard very highly, invites the listener to reflect on the relationship between concrete and abstract, between noise, sound and instrumental music.

V

Similar to aeroplanes, rockets and spacecraft have challenged the imagination of contemporary observers. Space engineers pursued their goals with religious zest;[52] stronger than aeroplanes, rockets served as symbols of Christ's ascension.[53] By means of rockets, man aimed at ruling the universe completely with space engineers as 'subcreators'.[54] The 'space race' from the late 1950s onwards was reflected in music. Shortly after Sputnik's launch on 4 October 1957, Duke Ellington wrote an essay on 'The Race for Space'.[55] It begins as a meditation on creation and the common roots of music and science and recognizes achievements such as 'Sputnik' as great works of both technology and art. Ellington links this to his theme of space and race, calculating the cost of racial prejudice in terms of lost creativity. According to him, racial prejudice resulted in the United States' failure to get a satellite into space first. But conditions in Jazz are different: in his conclusion Ellington regards jazz groups as microcosmic utopias, reflecting the spirit of harmony and creativity:

> 'So this is my view of the race of space. We'll never get it until we Americans, collectively and individually, get us a new sound. A new sound of harmony, brotherly love, common respect and consideration for the dignity and freedom of man.'.

From the late 1950s the jazz musician Sun Ra wrote pieces like 'We Travel the Spaceways', 'Rocket Number Nine Takes Off for the Planet Venus', 'Space is the Place' or 'Interplanetary Music', referring to imaginary and metaphorical images of space and space travel. They signify the composer's technological enthusiasm with its possiblity to escape from earth's gravity. This, in Sun Ra's sense, includes social restrictions and cultural and physical resistance to change. As in the blues pieces on railways, space travel is a metaphor for escape, rescue and freedom. For Sun Ra, technology is not opposed to magic. The space age means a final chance to go home, to climb to the ultimate mountaintop. For him, space is a metaphor of both exclusion and exterritorialization, of claiming the 'outside' as one's own, linking a revised and corrected past to a claimed future. Space serves as a metaphor to transvalue the dominant terms, to render them marginal while the outside world, the beyond, becomes central.[56] But in spite of racial differences Sun Ra viewed the success of the U.S. space program with admiration. Following the successful moon landing of Apollo 11 on 20 July 1969 he wrote his 'Walking on the Moon', dedicating it to the astronaut Neil Armstrong.

Sun Ra in a promotional still for *Space is the Place* in 1973 (© 1973 Jim Newman), John F. Szwed, *Space is the Place. The Lives and Times of Sun Ra* (New York, 1997), between 206 and 207.

Space is also a topic of 'art music'. In his dramma-oratore (Ceux de l'Espace), Vladimir Vogel deals with the theme of space travel. This work from 1970, premiered in 1979, deals with 'Notations after Leonardo da Vinci', 'Imaginations after Jules Verne' and 'Signals and Indications of Cosmonauts' and is a demanding and comprehensive treatment of a theme which has attracted a great deal of media attention during the last forty years.[57]

VI

Trains and planes as a theme in music? How do they enhance our understanding of the history of technology, the history of music, or both? If one's intention is to say something meaningful about the relationship between technology and culture, this kind of source material must be of prime importance.

With early industrialization, technology and transportation became important themes in music. Composers were intrigued by the power, speed and energy of locomotives or aircraft, physical attributes which they translated into driving rhythms and insistent ostinatos. They associated ideas like liberation from time and space with it and the reception of this music differed according to race or gender.

The idea that technology is reflected in culture and that culture influences technology, while certainly true, amounts to little more than a cliché. Yet a deeper understanding of this mutual influence demands speculation. As far as railways and aeroplanes as themes in music are concerned it is safe to say that most composers viewed them in a positive light. When those new technologies – railways in the first half of the 19th century, aircraft in the late 19th and early 20th centuries, rockets and spacecraft in the late 1950s and early 1960s – were implemented, composers wrote pieces dealing with them. As these themes were *en vogue*, compositions usually appeared in bunches.

Most of those compositions reflect the musicians' fascination with railway engines, aircraft and rockets. In times of war, especially during the Second World War, composers, commissioned by the Air Force, wrote music in praise of pilots and aeroplanes. But sometimes, and not only in times of war, technology is shown in a less positive light, as in works that depict crashes or other disasters.

In music, however, a positive image of technology generally prevails. In this, literature or the visual arts, especially paintings, differ from music: the former deal with the dangerous aspects and destructive uses of transport technology more prominently and distribute positive and negative aspects more evenly.[58]

Why then, is music different? Probably because of the aesthetic conditions of the artistic genre: to compose a piece consisting mainly of dissonances would appeal neither to musicians nor audiences. Although composers like Ornstein and Antheil did not hesitate to shock their audience with their musical language, their message was not critical of technology. Skeptical composers would probably have avoided the topic altogether. Many musicians of the first half of the 20th century wrote film music on technological themes and, when writing 'art music', were quick to point out that this was not intended to be 'programme music'.

Although the findings of this article have, hopefully, shed some light on the depiction of technology in music, there is still much room for detailed investigations into this important aspect of the relationship between technology and culture.

Notes:

1 G. Antheil, *Bad Boy of Music* (New York, 1975), 7. I should like to thank Paul McCutcheon, National Air and Space Museum, Washington D.C., for guiding me through the Landauer Collection of aviation songs, Dan Morgenstern and Ed Berger, Institute of Jazz Studies, Newark, New Jersey, Ulrich Duve, Klaus-Kuhnke-Archiv für populäre Musik, Bremen, with its incredibly rich stock of sound material, Dr. Wolfram Knauer, Jazz Institut Darmstadt, the archivists of the National Museum of American History, Washington D.C., with its enormous collection of songs on transport, and Roger Bilstein for valuable hints.

2 J. Bossin 'Strawinsky-Antheil-Maschinenmusik. Strawinsky und die neue Ästhetik des Antipathos, *Absolut modern sein: Zwischen Fahrrad und Fließband-culture technique in Frankreich 1899-1937*, ed. Neue Gesellschaft für Bildende Kunst (Berlin, 1986): 238-242.

3 F. K. Prieberg, *Musica ex Machina. Über das Verhältnis von Musik und Technik* (Berlin, 1960). E. Mayer-Rosa, *Musik und Technik. Vom Futurismus bis zur Elektronik* (Wolfenbüttel, Zürich, 1974). H.-J. Braun, 'Technik im Spiegel der Musik des frühen 20. Jahrhunderts', *Technikgeschichte*, 1992, 59: 109-131.

4 C.-H. Mahling, 'Musik und Eisenbahn. Beziehungen zwischen Kunst und Technik im 19. und 20. Jahrhundert', in *Studien zur Musikgeschichte. Eine Festschrift für Ludwig Finscher*, ed. A. Laubenthal and K. Kusan-Windweh (Kassel, 1995), 539-559.

5 P. Collaer, *Geschichte der modernen Musik* (Stuttgart, 1963), 489-490.

6 F. Sonnenberger, 'Mensch und Maschine. Technikfurcht und Techniklob am Beispiel Eisenbahn' in *Zug der Zeit – Zeit der Züge. Deutsche Eisenbahnen 1835-1985*, ed. Eisenbahnjahr Ausstellungsgesellschaft mbH, Nürnberg, 2 vols. (Berlin 1985), vol. 1: 24-37.

7 S. Kern, *The Culture of Time and Space 1880-1918* (Cambridge, MA, 1983), 123.

8 R. U. Ringger, 'Technik und Sport bei Arthur Honegger', in *Von Debussy bis Henze. Zur Musik unseres Jahrhunderts*, ed. R. U. Ringger (Munich, 1986), 16-23. See also A. Honegger, *Ich bin Komponist* (Zürich, 1952), 115-6

9 P. Oliver, *Screening the Blues. Aspects of the Blues Tradition* (London, 1968), 213; P. Oliver, *Blues Fell This Morning. Meaning in the Blues* (Cambridge, 1960), 58

10 G. Schuller, *The Swing Era. The Development of Jazz 1930-1945* (Oxford, New York, 1989), 62-65; J. E. Hasse, *Beyond Category. The Life and Genius of Duke Ellington* (New York 1993), 190.

11 P. Bussy, *Kraftwerk – Man, Machine and Music* (Wembley, Middx.), 87; T. Hoesch, 'Kraftwerk-Menschenmaschine', *Keyboards*, May 1997: 56-63.

12 J. Coltrane, The Africa Brass Session Vol. 2: The John Coltrane Quartet, Africa Brass Volumes 1&2, Impulse CD MCAD-42001.

13 B. Hoffmann, 'Zug um Zug – das Train-Motiv in der afro-amerikanischen Musik', in *'Und der Jazz ist nicht von Dauer' Aspekte afro-amerikanischer Musik. Festschrift für Alfons Michael Dauer* ed. B.Hoffmann and H. Rösing (Karben, 1988), 175-204: 181.

14 R. Reitz, 'Sorry, But I Can't Take You. Women's Railroad Blues', Covertext of Rosetta Records RR 1301, 1980.

15 A. Y. Davis, *Blues Legacies and Black Feminism. Gertrude 'Ma' Rainey, Bessie Smith, and Billie Holiday* (New York, 1998), 74

16 D. Hajdu, *Lush Life. A Biography of Billy Strayhorn* (New York, 1996), 85; Schuller (see note 10), 136; Hasse (see note 10), 268.

17 Mahling (see note 4), 546.

18 See R. E. Rodda's Covertext, 'Villa-Lobos, Bachianas Brasileras' nos. 2, 4 & 8', TELARC CD-80393.

19 See Hasse (note 10), 300.

20 R. Frisius, 'Konkrete Musik. Ein Lehrpfad durch die Welt der Klänge', *Neue Zeitschrift für Musik*, Sept./Okt. 1997: 14-21; P. Schaeffer, *Musique concrète. Von den Pariser Anfängen um 1948 bis zur elektronischen Musik von heute* (Stuttgart, 1974).

21 P. Manning, *Electronic and Computer Music*, 2nd ed. (Oxford, 1993), 20.

22 Frisius, (see note 20), 16.

23 R. Fanselau, 'Steve Reich', in *Komponisten der Gegenwart*, ed. H.-W. Heister and W.-W. Sparrer (Munich, 1992).

24 Mahling (see note 4), 543-4

25 Schuller (see note 10), 61-62.

26 For this and the following see N. Cohen, *Long Steel Rail. The Railroad in American Folksong* (Urbana, Chicago, London, 1981).

27 R. Wohl, *A Passion for Wings. Aviation and the Western Imagination 1908-1918* (New Haven, London, 1994), 1-2.

28 J. J. Corn, *The Winged Gospel. America's Romance with Aviation, 1900-1950*, (New York, 1983), VII.

29 M. S. Sherry, *The Rise of American Air Power. The Creation of Armageddon* (New Haven, 1987), 2.

30 M. Paris, *From the Wright Brothers to Top Gun. Aviation, Nationalism and Popular Cinema* (Manchester, New York, 1995), 4-8.

31 P. Fritzsche, *A Nation of Flyers. German Aviation and the Popular Imagination* (Cambridge MA, London, 1992), 2.

32 H. de la Motte-Haber, 'Musik aus der Maschine', in *Die Mechanik in den Künsten. Studien zur ästhetischen Bedeutung von Naturwissenschaft und Technologie*, ed. H. Möbius and J. J. Berns (Marburg, 1990): 223-230, 225.

33 J. Noller, 'Francesco Balilla Pratella' in *Komponisten der Gegenwart*, ed. H.-W. Heister and W.-W. Sparrer (Munich, 1992).

34 C.A. Tisdall and A. Bozzola, *Futurism* (London, 1977), 117.

35 V. Perlis, 'The Futurist Music of Leo Ornstein', *Notes. The Quarterly Journal of the Music Library Association 1975*, 31, No.4: 731-750, 741-2.

36 There are numerous examples of this in the Bella Landauer Collection (National Air and Space Museum Washington, D.C.).

37 L. Whitesitt, Covertext 'George Antheil. Bad Boy of Music', Troy CD 146. G. Antheil, *Bad Boy of Music* (Hollywood, 1960), 22.

38 L. Whitesitt, *The Life and the Music of George Antheil 1900-1959*, (Ann Arbor, MI, 1983), 88.

39 E. Townley, *Tell Your Story: A Dictionary of Mainstream Jazz and Blues Recordings, 1951-1975* (Chigwell, 1987).

40 G. Kleinen, 'Entwurf radiophonischer Musik in den Zwanziger Jahren', in *Radiophonische Musik*, ed. G. Batel, G. Kleinen and D. Salbert (Celle, 1985), 15-54, 18-19; S. Zielinski, *Audiovisionen. Kino und Fernsehen als Zwischenspiele in der Geschichte* (Reinbek, 1989), 118; G. Schubert, *Paul Hindemith* (Reinbek, 1981), 73-4; R. Sanders, *Kurt Weill* (Munich, 1980), 148-9.

41 M. E. Cory, 'Soundplay: The Polyphonous Tradition of German Radio Art', in *Wireless Imagination. Sound, Radio, and the Avant-Garde*, ed. D. Kahn and G. Whitehead (Cambridge, MA, London, 1994): 331-371.

42 Braun (see note 3), 116.

43 T. Hirsbrunner and P. Griffith, Covertext 'Honegger', CD D 435438-2.

44 M. Safranek, *Bohuslav Martinu. Leben und Werk* (Kassel, 1964), 123; H. Halbreich, *Bohuslav Martinu. Werkverzeichnis, Dokumentation und Biographie* (Zürich, 1968), 222; W. Green, *Famous Fighters of the Second World War* (London, 1957), 84-91.

45 N. Broder, *Samuel Barber* (New York, 1954), 36-37.

46 B.B. Heyman, *Samuel Barber, The Composer and His Music* (New York, Oxford, 1992), 219-223.

47 N. Broder, (see note 45), 81.

48 B.B. Heymann, (see note 46), 230.

49 See the material on Marc Blitzstein's Airborne Symphony in the Bella Landauer Collection, National Air and Space Museum, Washington D.C. See also the covertext by S. Ledbetter, Marc Blitzstein, Symphony: 'The Airborne', CD 'Leonard Bernstein, The Early Years III', RCA Victor 09026-62568-2

50 By Eddie Seiler, Sol Marces and Fred Jay, 1944.

51 D. Gutknecht, 'Stockhausen: Helikopter-Streichquartett. Uraufführung am 28.7., Salzburg, Mozarteum', *Österreichische Musik-Zeitschrift*, 1994, 49: 466-7.

52 See T. Crouch, *The Bishop's Boys* (New York, 1989), 33.

53 D. F. Noble, *The Religion of Technology. The Divinity of Man and the Spirit of Invention* (New York, 1997).

54 W. McDougall, *The Heavens and the Earth: A Political History of the Space Age* (New York, 1985), 4.

55 This essay remained unpublished. See J.F. Szwed, *Space is the Place: The Lives and Times of Sun Ra* (New York, 1997), 139.

56 Szwed (see note 55), 140.

57 H. Oesch, 'Wladimir Vogel' in *Komponisten der Gegenwart*, ed. H.-W. Heister and W.-W. Sparrer (Munich, 1992).

58 H.-J. Braun and W. Kaiser, *Energiewirtschaft, Automatisierung, Information seit 1914*. Propyläen Technikgeschichte, Vol. 5 (Berlin, 1992): 255-279.

Karin Bijsterveld

A SERVILE IMITATION. DISPUTES ABOUT MACHINES IN MUSIC, 1910-1930

1. *Introduction*[1]

In 1927, George Antheil's 'Ballet Mécanique' had its American première in Carnegie Hall, New York – featuring ten pianos, a pianola, xylophones, electric bells, sirens, airplane-propellers and percussion. The concert of 1927 had been advertised by referring to the riotous first performances of 'Ballet Mécanique' in Paris the year before, and Antheil himself had been introduced as a 'sensational American modernist composer'. However, the newspaper critics expressed a deep disappointment. New York did neither appreciate the pompous publicity nor the performance of 'Ballet Mécanique' itself. 'Mountain of Noise out of an Antheil' was one headline. 'Boos Greet Antheil Ballet of Machines' was another.[2]

The concert as well as the rumour by which it was surrounded had quite a lot in common with the first public performance of Luigi Russolo's 'Awakening of a City and Meeting of Automobiles and Airplanes' in Milan, 1914. This concert followed the publication of Russolo's Futurist manifesto 'The Art of Noises', in which Russolo proclaimed a renewal of music by enlarging and enriching the limited timbres of traditional orchestras with the infinite variety of timbres of noises. For his compositions, or 'networks of noises', Russolo used a complete orchestra of newly invented noise instruments. In 1914, 23 such instruments were on stage. A huge crowd had gathered, whistling, howling and throwing things even before the concert had started, and it remained in great uproar throughout the performance.[3]

Fig. 1: The 'intonarumori' (noise instruments) and their inventor Luigi Russolo with his friend Ugo Piatti (From Bartsch, 1986).

In those days, scandalous concert events surrounded with publicity like the ones of Russolo and Antheil were far from unique. In 1913, the first performance of Igor Stravinsky's 'Sacre du Printemps' in Paris caused such a noisy protest that the public itself became part of the performance. In Vienna, Arnold Schönberg's 'Kammersymphonie', again in 1913, was likewise accompanied by the shouting and rattling of front-door keys of concert visitors. Nor was the 'representation' of machines and machinery noise in music unique. Hector Berlioz composed 'The Singing of the Train' for choir and orchestra in 1846. Arthur Honegger, to give another of many existing examples, presented his 'Pacific 231' in 1923. Even the use of mechanical – automatic – instruments, such as the player piano, was far from new. Mozart had already composed for music automatons. A real novelty, however, was the construction of music out of everyday noise of machines with the help of machines or machine-like instruments instead of traditional orchestral instruments. Many of these instruments were invented in the first decades of the twentieth century – some of them being automatons, others creating new timbres and tones by making use of electricity and radiowaves.

The introduction of machines and machinery sound in music has often been discussed as being part of the legacy of the Futurist movement and its fascination for machines and the sound of everyday life. Although, generally speaking, this connection is correct, such statements mask the fact that composers of machine music often accused each other of being mere imitators of urban sound, thus strangely reproaching their colleagues for staying too close to their common denominator: the reverence for machines and everyday noise. The aim of this paper is to explain this non-fitting fact by showing how artists fostering machine music attributed distinct, even opposing, characteristics to machines, and how those qualifications were connected with their views on nature and – even more important – with their ideas concerning the immutable essence of life.

This will be illustrated by discussing the publications of Russolo and Antheil, as well as those of artists and critics, especially Mondriaan, reflecting upon the music and musical instruments of those two.[4] Their machine art did certainly not stand alone – Alexander Mosolov's 'Steel Foundry' and Sergej Prokofiev's 'Pas d'Acier', both dating from 1927, are other examples – and is assumed to have passed along to the compositions of Edgar Varèse, Pierre Schaeffer and John Cage. This paper concentrates on the ideas of and comments on Russolo and Antheil, however, because both made machine art out of machines, and because their works were fiercefully debated in the art and music literature of their time. Moreover, since much about them is already well-known[5], this article focuses on those aspects of their views that need to be explicated in order to understand the accusations of imitation.

2. *Machinery noise in music: embracing the dynamics of modern life*

The question as to how machinery noise became part of music is usually answered by referring to the programme of the Italian Futurist movement. Textbooks on twentieth century music delineate Futurism as the 'first clear manifestation' of 'a major and enduring concern' for 'the relationship between new music and modern technology'.[6] According to this view, the experimental nature of Futurism inspired numerous composers 'to capture the dynamics of the machine', 'to extend the resources of sound', and to express the 'love of mechanized urban life'.[7]

Fillipo Tommaso Marinetti and other Futurists indeed wanted art to break with the past and follow the dynamics of modern times. According to them, new forms of technology, such as the telephone, gramophone, movie, train, automobile, zeppelin and airplane, had deeply influenced human experience, resulting in a complete renewal of sensibility.[8] Since 'order, meditation and silence' had disappeared from life, art should likewise change.[9] 'Life' itself, and most of all 'movement' and 'energy' were the values Futurists wished to express.

These values, however, did not only reflect 'Bruitism' – the partiality of the Futurists towards the daily phenomena of modern existence – but *also* a tendency to express a semi-spiritual 'life force'. As the painter Gino Severini made clear in *De Stijl*, beauty was to be found in 'universal movement', in 'life'. Therefore, subjects such as motorcars, fast trains or dancers were preferred, because they enabled artists to express a maximum of reality. Such subjects, however, were certainly not necessary. A trotting horse likewise contained 'universal movement'.[10] As will be discussed below, this aspect of Futurism – which is often mentioned by art historians but seems to be absent from the textbook image of Futurist music – is important in order to understand why composers inspired by the Futurist machine cult often criticized each other's music.

Considering the engagement of Futurism with man's new experiences and life itself, Russolo's proposal to enrich compositions with noise was an extension of the ideas of Futurist writers and painters into the domain of music. According to Russolo, today's noise was 'triumphant' and reigned 'sovereign over the sensibility of men'.[11] Machines, as Russolo put it, had created such a variety of sounds that traditional tones alone, given their evenness and monotony, did not arouse enough emotion any more. To affect the modern listener required more than diatonism and chromatism. Hence, the timbres of noise, including those of modern war, should be added to the sound of traditional instruments.

Russolo's stress on enlarging and enriching, not completely substituting traditional sound, explains the fact that his noise instruments or *intonarumori* not only contained the acoustical phenomena of new forms of technology, but also those of nature. The 'bursters', for instance, produced a noise like that of 'an early automobile engine', the 'hummers' resembled the sound of 'an electric motor or the dynamos of electric power plants' and the 'howlers' were like sirens. The 'hissers', however, sounded like heavy rain, 'whistlers' like wind and 'croakers' like frogs.[12] Russolo's wish to enrich, not to replace, traditional sound also makes it understandable why he recommended the aesthetical quality of machinery and city noises by comparing these sounds with the noise of nature, which was traditionally considered to be pleasant, like 'the roaring of a waterfall' or 'the gurgling of a brook'.[13]

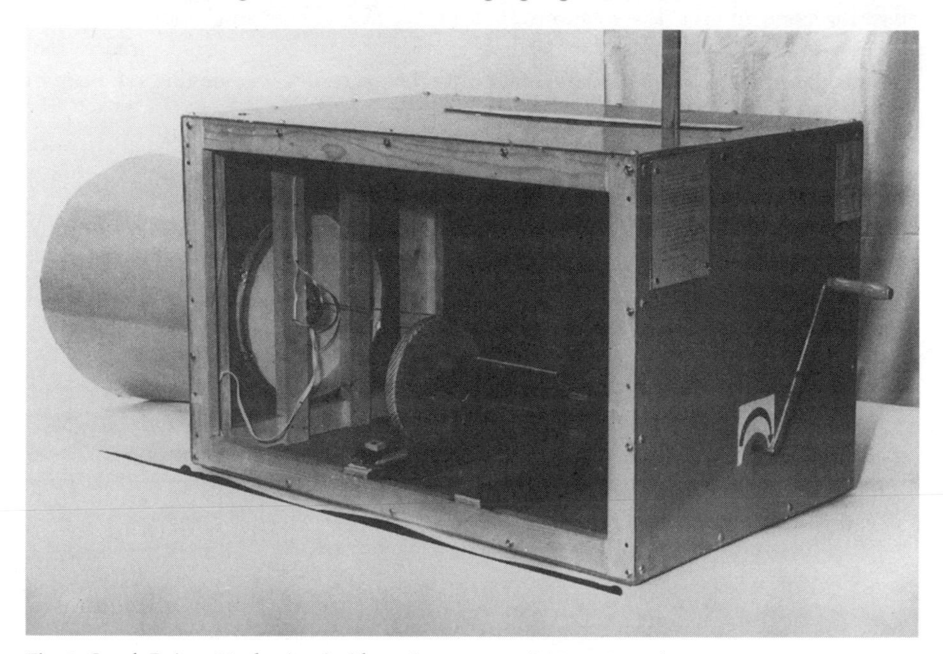

Fig. 2: Crank Driven Mechanism inside an 'intonarumori' (From Bartsch 1986: 15).

At the same time, however, both Russolo and Marinetti defended the bruitist experiments against the apparently expected accusation of being a mere 'imitation' of surrounding noise. 'The (...) noise networks are not simple impressionistic reproductions of the life that surrounds us', Marinetti said, 'but moving hypotheses of noise music. By a knowledgeable variation of the whole, the noises lose their episodic, accidental, and imitative character to achieve the abstract elements of art'.[14] Russolo was likewise aware of the dangers threatening his music: '(...) The Art of Noises would certainly not limit itself to an impressionistic and fragmentary reproduction of the noises of life. Thus, the ear must hear these noises mastered, servile, completely controlled, conquered and constrained to become elements of art.' Only by mastering the infinite complexity of noise, 'multiplying a hundredfold the rhythm of our life', by stirring the senses 'with the unexpected, the mysterious, the unknown', would one truly be able to move the soul.[15]

Russolo, who started his career as a painter and only later shifted his activities to music, was not the first Futurist announcing new forms of music. Between 1910 and 1912, Ballila Pratella published several manifestos in which he turned against traditional forms, rhythms and tonality, and fostered the use of microtones.[16] Even before this, in 1907, Ferruccio Busoni had published his famous essay 'Outline of a New Aesthetic of Music'. Busoni wanted to do away with programme music, advocating absolute music: true creations, free from all material limitations. Such hindrances were the unalterable properties of traditional musical instruments. All possible steps had to be taken to create an 'abstract sound, a technique without hindrance, an unlimited world of tones'.[17] Busoni's new world of tones included new divisions of the scale and divisions of the whole tone into microtones. He therefore mentioned the 'Dynamophone' of Thaddeus Cahill, a kind of electric organ, which was able to produce Busoni's scales.

Russolo, much like Busoni, wanted to create music making use of all possible microtones – a subject filling two chapters of 'The Art of Noises'. For such a system of microtones Russolo used the word *enharmonicism*, as Pratella had done before him. Conventionally, this term referred to enharmonically equal tones, like E-sharp

Fig. 3: Score 'Risveglio di una citta'. (From Russolo (1986/1916): 72-73).

and F, or B-sharp and C. On instruments with a tempered system, such as the piano, these tones are really equal. On wind and string instruments, however, this is not the case. Russolo considered the tempered system to be inferior: it could be compared to a 'system of painting that abolishes all the infinite gradations of the seven colours' – knowing red, but no rose and scarlet lake.[18] Therefore, enharmonic instruments should be created, being able to change pitch by enharmonic gradations instead of diatonic or chromatic leaps in pitch. Just as the howling of the wind produced enharmonic scales and the even richer world of machine noise was constantly enharmonic in the rising and falling of its pitch, enharmonic instruments should do so likewise.

3. *Nature and the essentials of machines*

Russolo's innovations were extensively debated in the international art press. Some commentators showed discontent with the instruments, for instance because the differences between the varying noise groups were considered too small.[19] Others accused the Futurists of producing sheer cacophony, sensation rather than emotion. According to Nicholas Gatty, the Futurists were breaking the laws of nature, since tonality, based on natural overtones, was grounded in the physical facts of sound. And how could variation be achieved, when concord and dissonant were both abolished, and when there was no regular rhythm which gave melody its frame and music its form? 'It is very difficult to believe', Gatty concluded, 'that music composed of such materials, a series of negatives, can have any really emotional value, that it can affect us in any other way than by giving us nervous shocks of a sensational, physical order'.[20]

Some artists felt attracted by Futurist theory, but were disappointed with Futurist musical practice. Among them were the composers Edgar Varèse and George Antheil and the painter Piet Mondriaan. In view of their comments, the fears of Marinetti and Russolo to get accused of just reproducing everyday noise proved to be right. Varèse dreamt of new instruments like the ones of Russolo, but did not appreciate Futurist music, which he regarded as a servile imitation of daily life.[21] As will be illustrated below, Mondriaan and Antheil had the same feelings. To phrase this in a different way, they accused the Futurists of venerating everyday sounds of machines *per se*, whereas contemporary textbooks exactly qualify this veneration as the common premise of machines art composers. To understand this seeming paradox, one needs to realize that artists such as Mondriaan and Antheil could accuse the Futurists of mere imitation because they deeply disagreed on which characteristics of machines had real artistic importance. Though they, indeed, shared an interest in technology and everyday sounds, they had quite different views on the essential qualities of machines. In case of Mondriaan these disagreements were connected with different ideas about nature and its place in music. Even more important for understanding the artistic clashes, however, were the varying views of the Futurists, Mondriaan and Antheil – to be discussed below – with respect to the immutable essence of life itself.

Mondriaan's aesthetic ideal was *neo-plasticism*. Neo-plasticism advocated universality against individuality, activity against passivity, manliness against femininity, inner against outer, spirit against matter, openness against closedness – even though ultimately, in line with Mondriaan's theosophist convictions, a harmonious whole of these oppositions should be achieved in a new culture. According to Mondriaan the old plastic arts had been dominantly individual: round, physical, bowed, closed – imitating nature. New plasticism, in contrast, should be universal: abstract, internal, mathematical, exact – as could be seen from the straight lines of neo-plastic painting. Reality was not denied, but should be abstract-real: art should express the deepest imaginable, pure and immutable reality. Since neo-plasticism should be achieved in all domains of art, music should likewise be renewed. Therefore the old instruments, whose full, vibrating, echoing tones and timbres basically imitated man's voice, reminding the listener of waves and curves, should be replaced.

According to Mondriaan, Futurists such as Russolo had done a good job in supplementing these timbres of traditional instruments with those of noise. He himself had attended one of their concerts in Paris in 1921. However, the Futurists' experiments had not gone far enough. Though they intended to make music more objective by making it more real, they basically stayed too close to nature. Their music lacked abstractness. The names of their imitative instruments – hummers, croakers – sufficiently proved this to Mondriaan. The fact that the Parisian concerts presented the more traditional compositions of Russolo's brother Antonio instead of Luigi's networks of noises like 'Awaking of a City' may have strengthened this opinion.[22] Moreover, Mondriaan argued that noise instruments were not really different from traditional instruments, since many of them could still perform diatonic and chromatic melodies, and besides, their use of 'scale' showed regression to the continuous fusion of natural sound. Nor were the bruitist instruments able to stop sound abruptly.

Mondriaan, however, considered the latter to be a prerequisite of neo-plastic music: instruments should have to keep wavelength and vibration fixed and be able to break off a tone suddenly. To this end instruments using electricity, magnetism and mechanics were the most suited, for they excluded the human, individual touch and enabled the perfect determination of sound. Even jazz was further ahead than the bruitists: jazz was further removed from harmony and its percussion made a sudden intervention of intuition possible.

Futurist music, Mondriaan stressed, was insufficiently freed from 'repetition'. He considered repetition a characteristic of both nature and the machine, though the latter showed *accelerated* repetition and therefore – since Mondriaan equated speed in music with straight lines in painting – enhanced objectivity. Art, however, should *not* display natural 'repetition' or its counterpart 'symmetry'. On the contrary, art should be made up of non-symmetrical relations: in painting between colour (red, yellow, blue) and non-colour (white, black, grey), in music between tone (sound) and non-tone (noise). Furthermore, successing tones should be different in volume and character. Therefore Mondriaan preferred the sound of traffic, which included irregular contrasts, to the repetitive rhythm of the carillon.[23] Finally, rhythm should be tight, such as in the *Stijlproeven* of his friend and composer Jacob van Domselaer, with whom Mondriaan exchanged ideas about music.[24]

So, unlike Russolo, Mondriaan did not stress the enharmonic quality of machines and thus the ability of machines to produce *continuity*, but their capacity to *fix* and *break off* tones, in order to create a tension between tone (sound) and non-tone (noise). Whereas the enharmonicism of Russolo had its counterpart in the infinite gradations of the seven colours, the sound of Mondriaan had its counterpart in primary colours. Moreover, Russolo saw no problem in associating his noise networks with nature, since nature, just like technology, was part of modern life, and since comparisons with nature could justify his experiments, making the latter seem less strange. Mondriaan, on the contrary, did everything to disassociate art from plain nature – art should be abstract-real. As he once said, however, the machine *itself* did *not* embody the new culture he strove for. Abstract reality could only be expressed by machines capable of creating a rhythm both fast and irregular, in contrasting relations.

George Antheil, member of *De Stijl*, likewise kept aloof from Bruitism, though others sometimes called him a Futurist composer. With respect to his critical stance towards musical Futurism he resembled Mondriaan. In addition, both were interested in the ability of machines to realise a fixed, precise rhythm.[25]

Although, as Van Dijk makes clear, both Mondriaan and Antheil were fascinated by 'movement' and 'time' – Mondriaan's counterpart of 'space' in painting – their ideas about music are less in line with each other than Van Dijk suggests. Whereas they may have had the same intention, namely, to produce abstract, streamlined, tense music, to achieve this, they had different means in mind. According to Mondriaan, old music tried to create contrasts by repetition and rests. Since such emptiness was immediately filled with the listeners' individuality, new music should constantly be expressive, by rapidly alternating tone (sound) and non-tone (noise).[26]

Antheil, however, made silences and repetition (of rhythm and by making 'loops') important elements of his 'Ballet Mécanique'. Time itself, he said, had to work like music. It was not tone (waves) or microtones that were important features of future music, but time. Antheil considered his 'Ballet Mécanique' as an example of the new fourth dimension of music, the first time-form. For him, time was the sole canvas of music, and not a by-product of tonality and tone. Time was inflexible, rigid and beautiful, the very stuff out of which life was made. According to Antheil, never a modest man, 'Ballet Mécanique' was the first work on earth composed out of and for machines, tonal nor atonal, just made of time and sound, without the traditional contrasts of piano and forte.[27]

In his autobiography 'Bad Boy of Music' Antheil explained that he had chosen the title for the 'brutal, contemporary, hard-boiled, symbolic of the spiritual exhaustion, the superathletic, non-sentimental period commencing »The Long Armistice«'. But he had certainly not meant the piece as 'a mundane piece of machinery'. Although he considered machines as very beautiful, he did not want to copy 'a machine directly down into music'.[28] Nor did he want to use machines like the ones the Italian Futurists used, since 'these had no mathematical dimension at all, nor claimed space, but just improvised noise that imitated motorcars, airplanes etcetera, which is ridiculous and had nothing to do with music.'[29]

Antheil's first version of 'Ballet Mécanique' had a score for sixteen pianolas, according to Whitesitt 'run electrically from a common control'.[30] 'Ballet Mécanique' was meant to accompany a movie, for which Fernand Léger was willing to collaborate. Yet the music became twice as long as the movie, and profound difficulties showed up in synchronizing the sixteen player-pianos. Therefore neither the collaboration with Léger (though the latter did make his movie), nor the pianola-version was brought to an end. As a consequence Antheil decided to orchestrate his ballet, using – apart from traditional pianos and percussion – one pianola, airplane-propellers, sirens and electric bells. After the debacle of Carnegie Hall, Antheil, still in his twenties, changed his style to neo-classicism, as Stravinsky had done. His reputation constantly hovered between that of a 'publicity-seeking rogue, charlatan and insincerist'[31] and a very talented, but not yet important composer.[32]

But he certainly secured himself another chapter in the history of machinery noise in music. Whereas Mondriaan felt attracted by essentials of machines different from Russolo, Antheil partly stressed other characteristics than Mondriaan. For Mondriaan, repetitive elements of machines were dangerous, reminding him of nature. So was silence – machines should instead account for a controlled *irregularity*. For Antheil, however, both silence and *repetition*, the latter beautifully produced by machines, was paramount to time-music. Moreover, neither Antheil nor Mondriaan were especially interested in microtones, which Russolo considered a fascinating quality of both nature and machines.

These diverging views on the principal qualities of machines make understandable why Mondriaan and Antheil both accused the Futurists of imitation, though the Futurists *themselves* were convinced of the abstraction of their noise networks. Since Mondriaan and Antheil chose other machine characteristics for artistic expression than Russolo, they did *merely* hear the noise of machines in the musical instruments and compositions of the latter.

Although Russolo, Mondriaan and Antheil unanimously stressed the capacity of machines to be *precise*, their beloved preciseness served different ambitions with respect to abstraction. Abstraction was to be found in the infinite, enharmonic complexity of noise (Russolo), the fast and irregular alternation of tone and non-tone (Mondriaan), or the mathematical time-form of repetition and silence (Antheil).

Moreover, these abstractions were far from non-committal, since such music had to make audible diverging conceptions of the essence of life: a maximized or multiplied reality (Russolo), an immutable universality (Mondriaan) or rigid, inflexible time – the very stuff out of which life was made (Antheil). These all-embracing, basic and constant values once more explain why Mondriaan and Antheil did not appreciate the works of the Futurists. Their machines had to express other ontologies than those of Russolo.

4. *The Man-Machine Debate in Music*

But the main debate about machines in music had yet to come. In 1924, the composer and critic Hans Stuckenschmidt prophesied that the traditional orchestra would be swept away in less than fifty years, to be replaced by new, 'mechanical' (which means 'automatical') instruments. Such a prediction was considered blasphemic in the world of classical music. Therefore, especially between 1926 and 1931, many journals published special issues about music and machines. Besides, the famous music festival of Donaueschingen devoted its 1926 meeting partly to mechanical music, inviting several composers to write for mechanical instruments and enabling inventors to demonstrate new instruments.

How did Stuckenschmidt support his shocking and widely published predictions? *Firstly*, he argued that, financially speaking, the symphonic orchestra depended heavily on the state and the city as well as on private fundings, since concerts were poorly visited and budget deficits grew. Because of this increase of financial problems, orchestras were in the long run condemned to death. *Secondly*, scores became increasingly complex, driving musicians to despair. Some works had already become, at least partly, unperformable: the intentions of composers exceeded human capacities of interpretation. *Thirdly*, individualism had passed and the new age had given place to a more collective attitude. Consequently, 'objectivity' was to be the new requirement of new music which, once the limitations of instruments and mistakes of performers had been removed – out of tune violins, the clattering of the wind instruments' valves, hesitant horn-players, oboists being out of breath – would make use of new forms of expression. Examples of these were new combinations of tones such as mathematically perfect intervals demanded by the modern twelve tone system, or new divisions of tones (quarter- or other microtones). Other new possibilities could be found in complex rhythms, with speed and force being augmented considerably, and in playing tied chords in fast tempo or in simultaneously playing several timbres with one hand. Moreover, from the fact that it was impossible to compose precisely, except for tempo and dynamics, it necessarily followed that one should search for the definite consolidation of a performance by means of notation.

Such complex and precise music could only be performed by machines: 'The machine', Stuckenschmidt said, 'has no limitations. Its strength and speed are practically limitless, its performance is of unfailing precision and uniform objectivity.'[33] New machine-like instruments should have a combination of the characteristics of the orchestrion and the electric or pneumatic piano, since music stamped directly on the record (paper roll) produced music with ideal precision and without an intermediate individual interpretation. Even phonographs and gramophones had a future as instruments, since they could create any tone-colour or rhythm one chose, even those non-existent in modern orchestras. Stuckenschmidt, like Busoni, also mentioned Thaddeus Cahill's 'Dynamophone', which by means of electrical vibrations could provide all conceivable timbres with any desired pitch, tempo and arrangement.

Thus, such new instruments were not only meant to solve the problems of the orchestras (since mechanical instruments were cheaper), and of the overworked musicians (since machines could take over the most complex music), but they would solve the problems of the age as well: striving for objectivity which was hindered by imperfect notation-systems and instruments. Stuckenschmidt did expect opposition. Without the human interpreter, critics said, the phonograph was inconceivable. Against this, Stuckenschmidt argued that only a single performance was necessary and that difficult score parts could even be recorded at a lower tempo. In addition, it was possible to produce a record graphically, like an etching, without involving a human interpreter. Such music would be like 'dead mechanics' and 'soulless automatons', opposers might say. Stuckenschmidt, however, stressed that machines could perform just as well as human interpreters. And, he continued, had not Stravinsky arranged 'Les Noces' and 'Sacre du Printemps' for the mechanical Pleyela-Piano as early as 1920 and had not George Antheil conceived his 'Ballet Mécanique' for mechanical interpretation? Musicians, Stuckenschmidt admitted, would indeed be deprived of their daily bread, but not before fifty years had passed.[34]

In short, Stuckenschmidt emphasized the machines' potential to cope with *complexity*, and to enable *preciseness*, a quality also stressed by Mondriaan and Antheil. Moreover, Stuckenschmidt admired the ability of machines to create the 'objectivity' that could perfectly match the collective attitude of his age – yet another view on the basic characteristic of life. With respect to the latter, many of Stuckenschmidt's 'progressive' colleagues agreed. The performance of music by machines was considered to proceed from the trend to compose a-sentimental, objective, mechanical music.[35]

But even the avant-garde audience of the festival of Donaueschingen had to get used to mechanical instruments. One day of the festival a Welte-Mignon (pianola) was put on, performing a piece by Ernst Toch. According to one of the listeners the instrument was a genius: unhumanly precise and clear. However, the audience hesitated, when the last tones died away. Should one applaud? For no living soul was on stage; just a machine. Finally, the public applauded, asked for a 'da capo' and behold: the Welte-Mignon did it again, exactly like it had done before.[36] Another visitor was less enthusiastic. He considered both the pieces by Ernst Toch and by Paul Hindemith, another 'invited' composer, as 'too cool and stiff'.[37]

Indeed, this was a much uttered argument against mechanical instruments,[38] as Stuckenschmidt had foreseen. Ernst Krenek knew that the reason for using machines in music was the need for clearness and coolness. But in the long run he felt that mechanical music was dull and dead. A jazzband performing complex polyrhythms with mechanical exactness created artistic excitement. But no such thing happened when a machine was used, for a machine was expected to be precise. Moreover, mechanical music left no room for the primitive drive for active participation.[39]

Willem Pijper, a Dutch composer, stressed that speed was a criterion which suited traffic, not music.[40] Though he himself had been impressed by the retorts of a gas factory in his native village, such emotions had to be considered as *a priori* extra-musical and should be sharply distinguished from the emotions induced by symphonies and sonatas.[41] Another Dutch commentator, the critic Herman Rutters, negatively evaluated the use of the siren as a musical instrument. He saw no use in machines overcoming technical problems, since he thought technical problems to be of artistic importance. The exertion needed to overcome the resistance of matter was paramount to artistic development.[42]

The most ironic comment came from the American Irving Weil. He did not like the music of Antheil and comparable composers. But perhaps Antheil was far along the main way. In the future the sense of hearing in the human animal would become thickened and muffled if the noise of daily life went on getting louder at the present rate of increase. In half a century or so man would be 'literally pig-eared, if not altogether as deaf as a moderately deaf post'.[43] The need for sufficient sound would therefore, among other things, lead to the use of noisemaking machines. A dreadful outlook for Irving Weil, but Russolo, Antheil, Mondriaan and Stuckenschmidt would, at least with respect to the predicted instruments, have welcomed it.

5. *Conclusions*

The use of machines and machinery sound in music was part of a broader movement of fascination for machines and the noises of mechanized, urban life. Textbooks of twentieth century music rightly stress that many artists followed the Futurists in capturing the dynamics of the machine and extending the resources of sound. However, these textbooks omit that at least part of these 'machine artists' also *followed* the Futurists in their effort to express some 'universal', semi-spiritual life force. And since they basically disagreed about exactly *which* machine characteristics could best express the deeper 'essence' of life's reality, they did not accept each other's use of machines as musical abstractions. Therefore Mondriaan and Antheil, two of the examples elaborated above, could blame their predecessor Russolo of merely imitating environmental noise, although their work and ideas had certainly been inspired by the Futurists. Thus, whereas such accusations of imitation do not fit the textbook image of machine music at first sight, they *do* fit, when the need to express the essence of life is brought in. *Without* knowledge of the latter, the accusations of imitation of machines would imply that the fascination for machines and every day sound was *not* the common denominator of the machine music movement after all. *With* knowledge of the need to express the ontology of reality one sees the source of Mondriaan's and Antheil's annoyance.

For making these essentials audible, artists stressed varying and often opposing characteristics of machines. *Continuity* (the production of microtones) and *discontinuity* (the sudden break off of tones), the ability to create controlled *irregularity* and controlled *repetition*, as well as the potentials for *precise complexity* were all considered to be the most important capacities of machines. Moreover, composers and other artists justified their viewpoints by aligning and opposing the concept of nature in different ways: different things being natural (enharmonicism, repetition, tonality) and to be sought for or to be avoided in music. Besides, critics of machine music did not always refute the love of machines, but argued against their specific association with music.

Russolo, Mondriaan, Antheil and Stuckenschmidt all used different formulations to make clear what they had in mind when looking for the essence of life, or, in the case of Stuckenschmidt, the basic feature of the age. Russolo wanted to maximize or multiply reality, Mondriaan tried to capture an immutable universality, Antheil had the intention of bringing time to the fore, and Stuckenschmidt strove for objectivity. Thus, the machine in music did not only function as a means to glorify the speed of modern times, but was simultaneously felt to vocalize reality's fundamentals, the real essence of life. Because *both* these issues were at stake, not the machine *itself*, but specific *characteristics* of the machine embodied the new culture.

Notes

1 I should like to thank Wiebe E. Bijker, Hans-Joachim Braun, Joke Spruyt, Manuel Stoffers, Jo Wachelder and Rein de Wilde for their helpful comments on earlier versions of this article.

2 L. Whitesitt, *The Life and Music of George Antheil, 1900-1959* (Ann Arbor, 1983), 31 and 37.

3 L. Russolo, *The Art of Noises* (New York, 1986/1916), 34.

4 Research into the reception of the music and ideas of Russolo and Antheil was done by focusing on the following journals (1910-1930): *Der Auftakt, Caecilia: Maandblad voor Muziek, Internationale Revue i10, Modern Music, The Musical Quarterly, Die Musik, Die Musikblätter des Anbruch, De Muziek, La Revue Musicale, De Stijl.*

5 See for instance: H.-J. Braun, 'Technik im Spiegel der Musik des frühen 20. Jahrhunderts', *Technikgeschichte*, 1992, 59: 109-131.

6 R.P. Morgan, *Twentieth Century Music. A History of Musical Style in Modern Europe and America* (New York, 1991), 117.

7 G. Watkins, *Soundings. Music in the Twentieth Century* (New York, 1988), 243-249.

8 M. Calvesi, *Der Futurismus. Kunst und Leben* (Köln, 1987), 11.

9 Marinetti quoted in G. Berghaus, *Futurism and Politics. Between Anarchist Rebellion and Fascist Reaction, 1909-1944* (Providence/Oxford, 1996), 17.

10 G. Severini 'Eenige denkbeelden over futurisme en cubisme'. *De Stijl*, 1919, 2: 25-27, 26 (Translation KB).

11 Russolo, op. cit. (note 3), 23.

12 B. Brown, 'Introduction' in: Russolo, op. cit. (note 3), 12.

13 Russolo, op. cit. (note 3), 25.

14 Marinetti, quoted in Brown, op. cit. (note 12), 18.

15 Russolo, op. cit. (note 3), 86-87.

16 I. Bartsch et al, *Russolo. Die Geräuschkunst, 1913-1931* (Bochum, 1986), 55.

17 Busoni, quoted in: A. Beaumont, *Busoni the Composer* (London, Boston, 1985), 90-91.

18 Russolo, op. cit. (note 3), 62.

19 A. Cœuroy, 'Les bruiteurs futuristes', *La Revue Musicale*, 1920/21, 1/2: 265.

20 N. C. Gatty, 'Futurism. A Series of Negatives', *The Musical Quarterly*, 1916, 2: 9-12 (11-12).

21 E. Varèse, 'Que la musique sonne', *Trois cent quatre-vingt-onze*, 1917, 5: 42.

22 K. von Maur, 'Mondrian und die Musik' in *Mondrian. Zeichnungen, Aquarelle, New Yorker Bilder* (Stuttgart, 1981), 289.

23 P. Mondriaan, P. 'Le neo-plasticisme', *De Stijl*, 1921, 4: 18-23; P. Mondriaan, 'De "bruiteurs futuristes italiens" en "het" nieuwe in de muziek', *De Stijl*, 1921, 4: 114-118, 130-136; P. Mondriaan, 'De jazz en de neo-plastiek', *Internationale Revue i10*, 1917, 1: 421-427; K. von Maur, *Vom Klang der Bilder. Die Musik in der Kunst des 20. Jahrhunderts* (München, 1985); Von Maur, op. cit. (note 22).

24 Mondriaan also talked about music with the architect J.J.P. Oud, the composer and critic Paul Sanders (see Y. Bois et al., *Piet Mondriaan* (New York, 1994), 36 and 42) and the composer Daniël Ruyneman. The latter created music with strong sound contrasts and new sound combinations (see Von Maur, op. cit. (note 23), 400). In 1919, Theo van Doesburg argued against the refinement of melody through the use of microtones and wrote in favour of music that brought harmonic structures into the forefront (see Von Maur, op. cit. (note 22), 291-292). Hardly anything, however, seems to be known about how these people exactly influenced Mondriaan's ideas about music.

25 M. van Dijk, 'Het schildersdoek van de muziek. Ballet mécanique van Fernand Léger en George Antheil', *Mens en Melodie*, 1996, 51: 58-62.

26 Mondriaan, 'De "bruiteurs"...', op. cit. (note 23), 133.

27 G. Antheil, 'Manifest der musico-mecanico', *De Stijl*, 1924, 6: 99-102; G. Antheil, 'Abstraktion und Zeit in der Musik', *De Stijl*, 1925, 6: 152-156; G. Antheil, 'My Ballet Mécanique', *De Stijl*, 1925, 6: 141-144.

28 G. Antheil, *Bad Boy of Music* (New York, 1981/1945), 139-140.

29 Antheil, 'Manifest', op. cit. (note 27), 101 – translation KB. Original text: '... die gar keine mathematischen Dimensionen hatten und auch keinen Anspruch auf Raum machten, sondern bloss Lärm improvisierten, die Automobile, Aéroplane usw. nachahmten, was lächerlich ist und mit Musik nichts zu tun hat.'

30 Whitesitt, op. cit. (note 2), 106.

31 R. Thompson, 'American Composers V. George Antheil', *Modern Music*, 1931, 8: 17-28, 20.

32 A. Copland, 'George Antheil', *Modern Music*, 1925, 2: 26-28; R. Petit, 'Ballet pour pleyela par George Antheil', *La Revue Musicale*, 1925, 7: 78-79.

33 H. H. Stuckenschmidt, 'Machines – A Vision on the Future', *Modern Music*, 1927, 4: 8-14, 9.

34 H. H. Stuckenschmidt, 'Die Mechanisierung der Musik', *Pult und Taktstock*, 1924, 2: 1-8; H. Stuckenschmidt, 'Mechanische Musik', *Der Auftakt*, 1926, 6: 170-173; H. Stuckenschmidt, 'Aeroplansonate' (George Antheil), *Der Auftakt*, 1926, 6: 178-181; H. H. Stuckenschmidt, 'Mechanisierung', *Anbruch*, 1926, 8: 345-346; H. H. Stuckenschmidt, 'Mechanical Music', *Der Kreis*, 1926, 3: 506-508; Stuckenschmidt, 'Machines', op. cit. (note 33).

35 K. Holl, 'Musik und Maschine', *Der Auftakt*, 1926, 6: 173-177; E. Steinhard, 'Donaueschingen: Mechanisches Musikfest', *Der Auftakt*, 1926, 6: 183-186; D. von Strassburg, 'Musikautomate', *Anbruch*, 1926, 8: 81-82; E. Toch, 'Musik für mechanische Instrumente', *Anbruch*, 1926, 8: 346-349.

36 Steinhard, op. cit. (note 35).

37 E. Felber, 'Step-children of music', *Modern Music*, 1926, 4: 31-33, 32.

38 A. Jemnitz, 'Antiphonie', *Anbruch*, 1926, 8: 350-353; 'Äusserungen über »Mechanische Musik« in Fach- und Tagesblättern,' *Anbruch*, 1926, 8: 401-404.

39 E. Krenek, 'Mechanisierung der Künste', *Internationale Revue i10*, 1927, 1: 376-380.

40 W. Pijper, 'Mechanische muziek', *Internationale Revue i10*, 1927, 1: 32-34.

41 W. Pijper, 'Pause del Silenzio', *De Muziek*, 1928, 2: 251-255/293-297.

42 H. Rutters, 'De sirene als muziekinstrument', *Caecilia. Maandblad voor Muziek*, 1913, 70: 1-8/A.

43 I. Weil, 'The Noise-Makers', *Modern Music*, 1927, 5: 24-28, 25.

Susan Schmidt Horning

FROM POLKA TO PUNK: GROWTH OF AN INDEPENDENT RECORDING STUDIO, 1934-1977

Historians have only begun to assess the impact of technology on the world of music and the art and practice of sound recording.[1] Since World War II, the popular music industry has grown exponentially, and along with it, the technology involved in making records. In an effort to examine the recording studio as a site of both technological and cultural change, this article traces the history of one independent studio in Cleveland, Ohio, over roughly a forty-year period. From its origins in radio broadcasting and transcription services to the production of multitrack rock albums, the Cleveland Recording Company spanned the technologies and the musical styles of two generations. It existed well before the postwar explosion of independent labels and recording studios, and remained active after many of these had closed their doors.[2] By tracing the work of the studio's chief engineer, Ken Hamann, we catch a glimpse into the nature of technological change at a time when standards of professionalization and equipment manufacturing had yet to be established. Like so many of the small studios and record labels that collectively helped to feed the burgeoning popular music industry of postwar America, Cleveland Recording provided a springboard for new and untried talent, and a kind of laboratory environment for technological and musical innovation.

In the mid-1930s, a Czechoslovakian emigré and radio announcer named Frederick C. Wolf started a modest recording studio in a downtown Cleveland office building with a Presto 16" transcription recorder. Exactly why Wolf decided to enter this field is somewhat of a mystery, since he was neither an engineer nor musician but primarily a businessman and 'politician by design.'[3] Gloria Busse Hamann began working for Fred Wolf before her future husband signed on as engineer in 1950. After impressing Wolf with her knowledge of music when she came into his studio to record a Christmas song for her family in 1943, Wolf hired Gloria on the spot.[4] The story she recalls hearing from her boss was that he had bought the recording equipment of two Cleveland businessmen, recently killed in a plane crash, in order to record his Sunday morning radio programs of live Czechoslovakian music over station WGAR. Wolf then capitalized on his investment by charging his fellow nationality program announcers for the same service; thus was born the Cleveland Recording Company.

In 1938, Wolf recorded the polka musician and band leader Frankie Yankovic when the future 'Polka King' and Grammy Award winner was still working in factories by day and playing in local taverns by night. After being turned down by both Columbia and RCA Victor for recording contracts, Yankovic recorded his first two 78s – three Slovenian polkas and a waltz medley – at Cleveland Recording. A pressing of 4,000 copies sold out within weeks from Anton Mervar's music store on St. Clair Avenue in the heart of downtown Cleveland. The following year Yankovic returned to Cleveland Recording to cut four more sides, and again they sold out almost immediately. In 1943, Yankovic joined the U.S. Army, but he continued to cut records at Cleveland Recording when home on furlough. Whether due to the limitations on his time or his pocketbook, the schedule was often rigorous. In those days, a typical recording session lasted three hours and yielded four songs.[6] But in his autobiography, Yankovic recalls one afternoon session during which his band cut thirty-two songs and sixteen records at Cleveland Recording:

> 'We had no time to fool around. If somebody hit a wrong note, we just kept going. One time one of the guys said we should record a song over. He said there were too many clinkers in it. I said, 'Leave the clinkers in. People like it better that way.'[7]

Like them or not, 'clinkers' were virtually unavoidable in these days before magnetic tape recording and the editing capabilities it introduced. Unless one could afford the time and expense of re-recording until everyone got it right, the initial performance, cut directly to a lacquer-coated aluminum disk, was recorded for posterity. Although attempts at magnetic recording dated back to the late nineteenth century, it was not until 1948 that the first professional music quality tape recorders became available to the American market, and it would be several years before all studios made the transition from disc to tape.[8] Yet, imperfect as they may have been, those early 78s got Yankovic signed to Columbia Records just after the war, and established Cleveland Recording's reputation as *the* place for Cleveland's polka musicians to record.

By the 1950s, thanks in part to the profits from Yankovic's recordings, Wolf began investing in more sophisticated equipment. After fulfilling a long-held ambition to open a radio station devoted primarily to nationality broadcasting, Wolf moved Cleveland Recording from its Huron Road location to the new studios of radio station WDOK on the fourth floor of the Loew's State Theater building at 1515 Euclid Avenue – a site which was 'absolutely suited for his purpose as a recording studio,' according to Wolf's colleague, radio announcer Wayne Mack.[9] Soon after he opened the station in 1950, Wolf hired a young navy veteran, Kenneth Richard Hamann, as an engineer for both WDOK and Cleveland Recording. He could not have made a better choice. Hamann had dabbled in electronics since childhood, studied at Case Institute of Technology, and attended the U.S. Navy Technical and Pilot Training Schools. Moreover, he held an F.C.C. First Class broadcast engineer's license and was an associate member of the Institute of Radio Engineers.[10] Over the next twenty years, Hamann's engineering expertise and technological innovations, coupled with Wolf's business sense and willingness to invest, made Cleveland Recording into the city's premier studio. More importantly, Hamann represented a new breed of recording engineer for whom creativity, experimentation, and adaptation were as essential to the job as skill in manipulating the controls.

Initially, Hamann was on staff at the radio station and one of a 'pool of engineers' Fred Wolf had hired to run recording sessions as they came in. Cleveland

Fig. 1: Ken Hamann at radio station WDOK, circa 1951. Note the RCA broadcast console with round control knobs, also used by Cleveland Recording Company.

Recording still operated on a part-time basis at this point, and most of the work served commercial clients: advertising agencies making commercial jingles, and corporations like Ohio Bell, Kirby Company, and Westinghouse who required synchronized recordings for the 35 mm strip films they used for employee training. Although on staff at WDOK and only 'moonlighting' in the studio, Hamann became increasingly drawn to the recording side, steeping himself in audio technology. As he recalls:

> 'What I did then naturally, without really thinking about it, was to begin to build in my own basement at home some equipment to add to the facility at Cleveland Recording. Eventually it got so complicated that Fred was very confused. So I guess, in a sense, I built in a kind of job insurance without really intending to, by making some of the equipment complex enough that I was the only one at the time who knew how to run it. But it *had* to be in order to do some of the things that we were getting into recording-wise.'[11]

As a teenager, Hamann had built a public address system and dual turntable set-up which he used to spin 78s at the 'Teenage Canteen,' a youth dance held weekly at the Lakewood Armory. Five years later, in an effort to overcome the limitations of disk recording, Hamann began building various monitoring and switching devices 'to improve the way that we could make records.' Soon, however, the studio converted from disk recording to tape recording, allowing Hamann's engineering skills the opportunity to expand. In 1952, Wolf purchased an Ampex 300 full-track monophonic tape recorder, a move which Hamann vividly remembers, 'opened up all kinds of vistas' for recording, editing, and sub-mixing.[12] Tape was still considered experimental in many studios. For years Capitol-Nashville continued to cut direct to disc, using an Ampex 200 tape recorder for backup until, according to engineer Jimmy Lockert, 'they had some confidence in the Ampex tape, and then they started using tape to make the record.'[13] Even the legendary Sam Phillips of Sun Records had only recently converted to tape at his Memphis Recording Service and would not acquire professional Ampex equipment until 1954.[14]

Throughout the 1950s, Hamann worked on developing stereo methods of recording as well as broadcasting. Like other hi-fi enthusiasts at the time, he experimented with 'ping-pong' stereo, recording the roller coaster at Euclid Beach Park and other environmental sounds on the Ampex 2-channel recorder the studio acquired in the mid-1950s. As early as 1954, he had begun stereo broadcasts over WDOK using both the AM and FM channels, and in October 1958, had devised and engineered the world's first 4-channel stereo broadcast, a live performance by the Dukes of Dixieland from Moe's Main Street over radio stations WERE-AM and FM and WDOK-AM and FM in Cleveland. Yet he still managed to find time for inventing; that same year, Hamann was awarded a patent for 'A New Binaural (Stereophonic) System of Disc Recording.'[15]

In 1959, impressed with the quality of the German 78s on Deutsche Grammophon they had received from Decca Records A & R man Lenny Joy, Wolf and Hamann went to Europe where they met with the German pro-audio manufacturer George Neumann, and visited recording sessions at Teldec and DG. They came back with the conviction that the Germans, as Hamann put it, 'were not doing anything we were not capable of doing, except that they were paying attention to detail – every little thing – and the results showed.'[16] Thus Cleveland Recording was among the first American studios to adopt German audio technology, then distributed exclusively through Gotham Audio in New York City. They had been using Neumann microphones since the early 1950s, but they now acquired a Neumann cutting lathe, and Hamann built a recording console based on the flat desk design he had seen for the first time in a Berlin beer hall that doubled as a recording studio during the week. Up to this point, most American studios still used the vertical broadcast console with round volume-control knobs; even Columbia Records' 30th Street Studio in New York maintained a recording console with knob controls as late as 1966.[17] As Hamann recalls: 'We were adapting broadcast equipment to recording; now it was time to do the opposite.'[18] So he found out where to purchase the necessary components, and designed and built a flat desk, 10-channel input, 3-channel output console employing vacuum tube technology. Rather than the control knobs, he installed linear faders that enabled him to adjust several channels with one hand during mixing, thus assuming a more interactive role in the recording process, and he incorporated equalization and limiting for each individual channel rather than the entire board.

Fig. 2: Three-channel mixing console with linear faders, designed and built by Ken Hamann (seated) about 1961 for Cleveland Recording Company.

By 1963, Hamann adapted 3-channel recording techniques to broadcasting. Calling it '3-D Radio,' Wolf and Hamann began a weekly 3-channel broadcast of live classical music from the nearby Oberlin and Baldwin-Wallace college campuses over the two channels of WDOK-FM stereo, and the single AM channel.[19] During his stereo experiments of the 1950s, Hamann had begun to focus on recording classical music. Loading his six foot high, twenty-two inch square rack of portable recording gear 'like a casket' into the back of a station wagon, Hamann recorded performances at the Oberlin College Contemporary Music Festivals conducted by Aaron Copland and Igor Stravinsky. These were exclusively remote recordings, but during this period the studio's clientele had begun to change, documenting a cultural shift taking place across America. Commercial clients and transcribed radio programs remained the primary source of Cleveland Recording's business, but the music clients now included not only polka groups, but popular singers like the Tracy Twins; occasional sessions by performers appearing in town like Andy Williams, Kay Thompson, and Mel Torme; and eventually, rhythm and blues and rock 'n' roll. Hamann recalls that 'some of the older musicians would come in and try to emulate some of the sounds they were hearing and they couldn't *do* it. They didn't have the *feel* for it. ... They simply could not emulate the sounds that the kids were getting.'[20] These sounds, and particularly the *volume* of rock 'n' roll, would put new demands on the technology of the recording studio and further drive Hamann's engineering innovations. And it was during this period that the interplay of music and technology, the symbiotic working relationship between musician and engineer that developed in the recording studio of the 1960s, began to emerge.[21]

Hamann's first rock 'n' roll recording session occurred shortly after he and Fred Wolf returned from Germany. Decca Records had commissioned Cleveland Recording to tape Bill Haley and the Comets' performance at the Masonic Auditorium. This experience not only proved to be a logistical nightmare for Hamann, who had to navigate all of the portable recording gear up a narrow flight of stairs into the auditorium's projection booth, it also taught him a few things. First, he was not prepared for the volume or intensity of rock 'n' roll, and neither were his microphones. The Neumann and Telefunken mics 'were very hot... the signals coming out of those microphones were overloading the input stages on the recorders,' and therefore causing distortion.[22] He had already learned that some of the louder passages of classical pieces distorted on tape, and he determined that the microphone inputs of the Ampex recorders were too sensitive for the Neumann U47 mics. His first remedy for this was to introduce plug-in equalizers between sections of the mic cable, but he eventually designed his own mixer in which he allowed for the discrepancy without having to introduce external attentuation. The other lesson he learned recording the Bill Haley concert was the wide variation in output volumes of the group's instruments and the ensuing difficulty of capturing drums and electric guitars as well as acoustic bass on tape. With only four microphones at his disposal, and one reserved for the show's announcer, disc jockey Bill Randle, Hamann learned that recording rock 'n' roll required techniques far different from those he had used recording a symphony orchestra.

By the 1960s, more and more rock groups came to Cleveland Recording, ranging from high school kids with parents willing to pay for studio time, to semi-professional bands wanting to record demos to send to record companies – groups like Joey and the Continentals and Rocco and the Flames. In 1964, Hamann recorded a rhythm and blues group, The Montclairs, whose song 'Happy Feet Time' was the first of the studio's records to hit the national charts in 1965. Around that time a local musician and band leader enlisted Hamann's aid in recording a song he felt was a sure hit. The resulting 3-channel recording of 'Time Won't Let Me' earned Tom King and the Starfires a contract with Capitol Records, a new name – The Outsiders – and the single rose to #5 on the *Billboard* pop charts.

That recording was significant for Hamann, not only because it was a huge hit, but because it represented hours of effort and experimentation to achieve the proper mix, and years of accumulated technique recording polka, jazz, classical, and now rock music. To reach the point of making a flawlessly balanced 3-channel recording of a song that featured drums, bass, guitars, organ, trumpet, saxophone, tambourine and vocals – all engineered so skillfully as to maintain crisp presence on the final record – Hamann had gone through what he calls 'an evolutionary process.' Some of the early recordings were, as he put it, 'generally speaking, pretty awful,' but each time he made a record he learned something and tried to build on the experience.[23] Not only did he have to devise ways to accommodate the louder drum and guitar sounds intrinsic to rock 'n' roll, as he had learned recording the Bill Haley concert, but the limitations of three channels forced him to anticipate the final product and to develop sub-mixing techniques to achieve the correct proportion of drums and bass on the first track, so that subsequent tracks would not overpower them.

By 1966, five years after Hamann had completed building his 3-channel mixing console, Cleveland Recording acquired a Studer 4-track recorder, which Hamann modified to 8-track two years later. This meant that the studio now had true multi-track recording capabilities; instruments now could be recorded more discreetly, and thus controlled more fully in the final mix. In July and October of 1967, Hamann recorded two more songs by local groups, both of which reached the *Billboard* 'Top Ten' during the week of February 3, 1968: The Lemon Pipers' 'Green Tambourine' and the Human Beinz' 'Nobody But Me.' A proliferation of rock groups recorded at Cleveland Recording in 1968, most notably the James Gang with Joe Walsh. The following year, a young producer/performer from Flint, Michigan, named Terry Knight, assembled the members of his former rock group, The Pack, to record four demos at Cleveland Recording. These sessions were the beginning of a long association between the studio and one of its most successful clients, Grand Funk Railroad.

After Knight secured Grand Funk a record deal with Capitol Records on the strength of their Cleveland Recording demos, they returned to Cleveland to record their first album. By then, Hamann had managed to talk Fred Wolf into buying a 16-track Ampex recorder with the proceeds from the sale of WDOK – a somewhat extravagant move which paid off.[24] Between 1969 and 1973, Grand Funk recorded a total of seven albums at Cleveland Recording, all of which quickly achieved gold record status.[25] With record sales like that, the group and its producer could have had their choice of studios. But Knight's allegiance to working at Cleveland Recording with Ken Hamann rested on both sound business sense and a keen appreciation for the studio's technological capabilities. When Don White, an engineer Knight had worked with at another Cleveland studio, tried to lure his former client back, Knight declined, explaining: 'When you're winning at poker, never get up from the table.'[26] Knight was also aware that he had found a well-kept secret. He told a *Billboard* magazine columnist in 1971 that the reason he worked only at Cleveland Recording was that: 'This place is a highly developed – as yet undiscovered – studio. ... I find it technically to be one of the five top studios in the country.'[27]

Ken Hamann was the technical wizard who had made this possible, but he credits Fred Wolf for having the innovative spirit that led him to bankroll the ventures, despite the fact that Wolf was 'all thumbs' in the studio to the extent that the engineers 'shuddered everytime he put his hands on certain pieces of equipment.'[28] Nevertheless, Wolf remained deeply involved in the studio until the late 1960s, when declining health forced him to retire from the studio business entirely. In 1970, Hamann and another engineer, John Hansen, bought Cleveland Recording and moved the studio to a new location, spending six months renovating an old Chevrolet dealership into an acoustical masterpiece. John Hansen focused on commercial clients, and Ken Hamann continued to record rock groups like Wild Cherry, the James Gang, Tiny Alice, and Brownsville Station, but the string of hit records he had engineered in the 1960s and early '70s was not to be repeated.[29] Musicians were notoriously poor paying customers, and it became increasingly evident, at least to Hansen, that the commercial clients kept Cleveland Recording afloat. Tension between the 'rather raucous' music clients and the 'rather straight-laced' commercial clients, as well as between Hamann and Hansen over money matters, finally led to what Hamann characterizes as 'a divorce' when, in 1977, the two owners decided to go their separate ways.[30] They divided the studio equipment, Hansen kept the name Cleveland Recording, and Hamann opened a new studio, Suma Recording, in an old country home east of Cleveland. Cleveland Recording continued in operation until Hansen's death in the early 1990s, and Suma Recording continues to record clients ranging from jingle writers to the pioneering alternative music group Pere Ubu.[31]

Throughout Cleveland Recording's most prolific recording activity, Ken Hamann continued to adapt and build equipment, including a massive 48-channel quadrophonic mixing console in 1972, but the 1960s also saw him devoting considerable energy to finding ways of accommodating the increasingly experimental recording ambitions of his music clients. One early example of Hamann's willingness to try anything, and of the extravagance that characterized much later rock studio work of the 1970s, was a session during which a rock group from Youngstown, Ohio, The Human Beinz, chopped up a piano in the basement of the Euclid Avenue building, while Ken miked the process, feeding the lines up to the fourth floor recording studio. In other sessions with the James Gang's Joe Walsh, who went on to greater renown first as a solo artist, then as member of the Eagles, Hamann recalls that 'We tried out different ways to modulate his voice. We ran his voice and/or guitar through a Leslie speaker, tried all kinds of different things, unusual things. Tried to get new sounds or interesting sounds that worked.'[32]

On the heels of the Beach Boys' 'Pet Sounds' and the Beatles' 'Sgt. Pepper' albums, many rock groups experimented to get sounds that were often unreproducible in a live performance. The goal was no longer simply to document a well-rehearsed performance, but to create new sounds, to layer and shape a song using the tools of recording technology. In the process, a new symbiotic working relationship had developed between musicians, engineers, and producers in the recording studios of the 1960s which had a profound impact on the music they made. As Hamann aptly sums it up:

> 'We were working together during those years, during the late '60s mostly, and it became much more of an art. I would say, in retrospect, a true art form, where the engineer and the record producer and the musicians were working together as a team to create music, sounds that portrayed some kind of image. Just like a painter... with colors on a palette. We were doing the same thing. And we were able to. We were told more than once that our *freedom*, the freedom that the producers and musicians enjoyed in our studio, was far beyond what they could have in

Fig. 3: 24/48-channel console at Cleveland Recording Company, 1974. Hamann designed the console for possible modifications, like computerized automated mixing, and included a TV monitor for recording for television.

> either New York, Nashville... and Los Angeles, in particular, because the unions out there would prevent anybody from getting their hands on the controls.'[33]

Like many independent studios of this era, Cleveland Recording acted as a springboard for new talent, offering a creative recording environment free of job-defined restrictions, and an adaptive and innovative approach to pushing the capabilities of available technology. Ken Hamann was not a musician, but his description of the recording process as 'a true art form' embodies a sentiment shared by his contemporaries. Atlantic Records producer and co-founder Jerry Wexler once described engineer Tom Dowd's skill at the mixing console in similar terms: 'Tom pushed those pots (the volume controls) like a painter sorting colors. He turned microphone placement into an art.'[34] In his study of sound mixers, sociologist Edward R. Kealy documented the changing professional identity of recording engineers over several decades. He concluded that by the early 1970s, three factors – technological developments, the decline of corporate and union control in studios, and the growth of a network of musical collaborators – had transformed their work 'from a technical craft to a modern art form.'[35]

In a business in which fame and fortune are motivating factors, engineers like Ken Hamann rarely received either recognition or adequate remuneration for their efforts. Instead, they were driven by the desire to improve and expand the technology and practice of recording, infused with a 'technological enthusiasm' for refining the tools of their craft and the integrity of the final product.[36] By creatively adapting old technology to new needs, updating their existing equipment to meet higher recording standards, and building or buying new equipment when it became available, their efforts suggest that technological change in the independent recording studios of the 1950s and '60s was contingent, multi-causal, and decentralized. Contingent, because it depended on the technical ingenuity of the engineer and his chance association with creative musicians, as well as the willingness of the studio owner to invest and take chances. Multi-causal, because the inspiration for change came from a variety of sources: the engineer, the musician, the producer, and the new demands placed on existing equipment by the volume and style of rock and roll. And decentralized, because these audio engineers worked largely within their own private realms, learning by trial and error. They comprised a scattered 'community of practitioners' who met annually at trade conventions like those of the Audio Engineering Society, sometimes trading their knowledge and experience, but more often closely guarding their hard-won secrets.[37]

By the late 1970s things had changed. The Society of Professional Audio Recording Services, established in 1979 by a select group of studio owners, included ownership of *at least* two 24-track recording machines as a qualification for membership. Recording had become big business and audio engineering had become an established field for which professional training or apprenticeship were increasingly becoming the entry requirements. Manufacturers of every type of recording component, from mixing consoles to outboard effects, made the self-taught tinkerer-engineer who built his studio from the ground up virtually obsolete. The state-of-the-art 48-track acoustically engineered recording studio with equalizers, harmonizers, digital delay, digital reverb, noise gates, limiters, expanders, tape duplicators and a full-time maintenance crew bore little resemblance to the disc recording studios of the 1930s, and required an initial investment that few could afford.[38] Within four decades, the field of sound recording had shifted from a broadly democratic small business opportunity to a highly specialized and competitive profession.

Cleveland Recording was among the few independent studios that managed to weather this transition. From the unavoidable 'clinkers' of Frank Yankovic's polka 78s to the multi-tracked synthesizers and screaming guitars of Pere Ubu's 'Final Solution,' its history vividly illustrates the technological and cultural transformation of studio recording in these years. In that time, technology had allowed new possibilities, and innovative engineers coupled with musicians eager to experiment, to break down the accepted standards of pop music convention in a musical climate that still encouraged individuality, ultimately resulted in new sounds, new standards, and a new working relationship between artist, producer, and recording engineer.

Notes

1. Until the 1990s, the best general histories of the phonograph and record industry were R. Gelatt, *The Fabulous Phonograph* (New York, rev. ed., 1965); C. Schicke, *Revolution in Sound* (New York, 1974); O. Read and W. Welch, *From Tin Foil to Stereo* (Indianapolis, 2d ed., 1976); and F. Gaisberg, *The Music Goes Round* (New York, 1977). Recent studies that examine the development of sound recording technology include S. Jones, *Rock Formation* (Newbury Park, 1992); S. Lubar, *InfoCulture* (Boston, 1993); A. Millard, *America on Record* (Cambridge, 1995); M. Chanan, *Repeated Takes* (London, 1995); J. Kraft, *Stage to Studio* (Baltimore, 1996); and M. Cunningham, *Good Vibrations* (Chessington, 1996).
2. Founded in 1934, Cleveland Recording Co. predated Chicago's Universal Recording Corporation (1946), Los Angeles' Gold Star Recording and the Sam Phillips' Memphis Recording Service (both 1950), to cite a few of the more well-known examples. See R. Pruter, *Doowop: The Chicago Scene* (Urbana IL, 1996), 15-17; W. Schwartau, 'Gold Star Recording: A Masterful Partnership,' *Mix: The Recording Industry Magazine*, 1982, 6:8, 30; and C. Escott with M. Hawkins, *Good Rockin' Tonight* (New York, 1991). Also, C. Gillett, *The Sound of the City* (London, rev. ed., 1983), is the best source for the proliferation of independent record labels from roughly 1942-1971.
3. Ken Hamann, interview with author, Painesville, Ohio, 13 November 1995.
4. Gloria Hamann, telephone conversation with author, 22 May 1996. Memento recordings like Gloria's Christmas record were one source of business for the early Cleveland Recording Company.
5. V. Greene, *A Passion for Polka* (Berkeley, 1992), 233-34; F. Yankovic, as told to R. Dolgan, *The Polka King* (Cleveland, 1977), 47.
6. T. Fox, *In the Groove* (New York, 1986), 36.
7. Yankovic, op. cit. (note 5), 67.
8. M. Clark, *The Magnetic Recording Industry, 1878-1960* (Ph.D. diss., University of Delaware, 1992), ch. 9.
9. Wayne Mack, telephone conversation with author, 11 May 1996.
10. Until the founding of the Audio Engineering Society in 1948, which Hamann joined shortly after being hired by Wolf, the IRE was virtually the only professional organization for audio engineers. See M. McMahon, *The Making of a Profession* (New York, 1984), 216.
11. Hamann interview, 13 November 1995.
12. Hamann interview, 13 November 1995.
13. Cited in D. Daley, 'Nashville's Original Engineers: Together Again,' *Mix*, 1996, 20:7, 96.
14. See Escott with Hawkins, op. cit. (note 2), 14-18.
15. U.S. Patent #2,849,540.
16. Hamann interview, 13 November 1995.
17. ' "Take One!" The Making of a Record: 2,' *Columbia Record*, February 1966, 8.
18. Hamann interview, 13 November 1995.
19. J. Frankel, 'Three-Dimension Broadcast Gives Cleveland Radio First,' *The Cleveland Press*, 17 April 1963.
20. Hamann interview, 13 November 1995.

21 For more on the collaborative nature of studio recording in this period, see E. Kealy, 'From Craft to Art: The Case of Sound Mixers and Popular Music,' *Sociology of Work and Occupations*, 1979, 6:1, 3-29.

22 Ken Hamann, interview with author, Painesville, Ohio, 25 October 1996.

23 Hamann interview, 21 November 1995.

24 Fred Wolf sold his interest in radio station WDOK in 1965.

25 In 1958, the Recording Industry Association of America established gold awards as the industry standard for success – sales of over half a million. In 1976, the RIAA awarded platinum to records selling over a million copies. By 1992 six of the seven GFR albums recorded in Cleveland had received platinum awards.

26 Don White, interview with author, Akron, Ohio, 23 January 1996.

27 Cited in C. Hall, 'Studio Track,' *Billboard*, 27 March 1971, 4. This column quotes Knight as referring to the 'Cleveland Sound Studios,' but this is a misquote.

28 Hamann interview, 21 November 1995.

29 Cleveland Recording's last gold records were Wild Cherry's single, 'Play That Funky Music,' and the subsequent album 'Wild Cherry,' recorded in 1976 and released on Epic/Sweet City Records.

30 Hamann interview, 21 November 1995.

31 Pere Ubu's recent CD box set release on the Geffen Records' DGC imprint, 'Datapanik in the Year Zero,' comprises material recorded at both Cleveland Recording Company and Suma Recording Studio. Nearly all of the group's thirty-five releases since 1975 have been recorded at Cleveland or Suma, and engineered by Ken Hamann or his son, Paul.

32 Hamann interview, 13 November 1995.

33 Hamann interview, 13 November 1995.

34 J. Wexler and D. Ritz, *Rhythm and the Blues* (New York, 1993), 82.

35 See Kealy, op. cit. (note 21), 25.

36 Much like the dragsters profiled in R. Post, *High Performance* (Baltimore, 1994).

37 I draw loosely on the notion of communities of technological practitioners as outlined in E. Constant, *The Origins of the Turbojet Revolution* (Baltimore, 1980).

38 In 1972, Hamann estimated that the equipment then used by Cleveland Recording, accumulated over the years by purchase or his own design, would have cost $400,000 to buy from scratch. M. Ward, 'Ken Hamann Seeks More from Artists' Gold Notes at Cleveland Studios,' *Plain Dealer Action Tab*, 29 December 1972.

ALEXANDER B. MAGOUN

THE ORIGINS OF THE 45-RPM RECORD AT RCA VICTOR, 1939-1948

The significance of the vinyl microgroove record playing at 45 revolutions per minute (rpm) for the production and consumption of music has traditionally paled before the attention paid to Columbia Records' longplaying record (the LP). This situation exists for historical and historiographical reasons. The RCA Victor division of the Radio Corporation of America (RCA) introduced the 45 seven months after Columbia brought out the LP in June 1948, leading writers to see it as a response to Columbia's coup. Because it offered no advantage in playing time over traditional 78-rpm records, critics considered the 45 a conservative innovation, dedicated to popular music. Playback on inexpensive record players and its predominant use for children's and pop records led to the assumption that the quality of sound reproduction at 45-rpm was inferior to that from an LP that played at 33 1/3-rpm. Since it was marketed at the beginnings of the revolution in pop music that led to rock 'n' roll, some writers have assumed that the format was as important to this phenomenon as magnetic recording and disc jockeys. Understanding of RCA's intentions has not been helped by the fact that the two standard histories of recorded sound have been written by classical music aficionados.[1]

One result has been to ignore or denigrate the role of the only major company in this period to produce both records and phonographs. Given RCA's industrial preeminence, its assumption of the Victor Talking Machine Company's trademarked reputation for quality, and its own extensive research into electronically transmitted sound, it is fair to ask why it chose to innovate a format that appeared to take the low road musically and aurally. This article proposes to answer that question. Closer study of the origins of the 45-rpm record, based on extant RCA files, modifies or corrects the traditional views. Despite the apparent conservatism of the 45, RCA Victor engineers and management had lofty expectations for their effort to standardise the phonograph record. Motivated by the expanding market for music in the late 1930s and the improvements in sound possible through the use of manmade plastic as a record medium, Victor devised a music delivery system that promised the consumer greater choice and the industry greater sales than ever before.

This article begins by examining the industrial, corporate and consumer contexts in which RCA Victor engineers began research and development on a new record in 1939. These initial stimuli gave way to a long gestation period before the 45 was introduced to the public ten years later. Innovation was complicated by debate over the nature of and desire for high fidelity, by the state of the economy and the record industry and by the competitive relationship between the leaders of RCA and the

Columbia Broadcasting System (CBS), corporate parent of Columbia Records. The article concludes with some consideration of the consequences of the 45 and the LP.

The stimulus to innovation in phonograph records at the end of the 1930s arose from developments in several technologies and the home entertainment industry. At its base was an awareness that the record was essentially the same one Eldridge Johnson perfected for mass production at the Victor Talking Machine Company at the turn of the century. Since then, however, recording and playback techniques had changed markedly. Electrical recording, popularised in the second half of the 1920s, and electrical playback, diffused in the 1930s, enabled potential reproduction of far greater fidelity to the original sound than discs designed for mechanically recorded and reproduced sounds could accomodate. Electrification and the competition of radio as a source of continuous entertainment also led to more intensive development of record-changing mechanisms.

Despite Victor's continuing marketing emphasis on quality of sound on its discs, the Depression hindered investment in improving record fidelity. Although RCA Victor had commissioned a study of recording and reproducing records for home consumption in the winter of 1932-33, most research and development focused on film, broadcast, and radio receiver technologies, where technical and commercial competition still existed.[2] As to the improvement of the record itself, again, early manmade substitutes for shellac were designed for the broadcast and soundtrack markets.[3] Vinylite in the 1930s was a prohibitively expensive replacement for the shellac and limestone filler used in popular discs.[4] Playback systems for the home enjoyed a trickle-down effect from these developments, but by the end of the 1930s consumers still played records containing the same ingredients of a generation before, with playback pickups still weighed in ounces, not grams.

As a result, despite marked improvements in recording technology, the best home phonographs and records of the late 1930s could still only reproduce a frequency range between 100 and 4,500 hertz (hz) and a dynamic range, the difference between the softest and loudest sounds, of 30 decibels (db).[5] Any attempt to offer higher fidelity to the mass market would require a complete overhaul of the process of recording and reproduction, and a redesign of the record and its player that incorporated and built on disparate strands of research.

By the end of the 1930s, however, the problems of sound quality were of minor importance to an industry enjoying sales totals rivaling those of the best pre-Depression years. RCA Victor officials in the Record and Home Instrument Departments could afford to be pleased with the success of their efforts to revive the phonograph record as a part of middle-class home entertainment.[6] The company's record sales more than tripled from 1932 to 1937, and then quintupled to an all-time high in 1941.[7] Production of record players rocketed from 6,000 in 1934 to 152,000 in 1939.[8] The most significant trend for the future, however, showed sales of radio-phonograph combinations rising from a one to a ten percent share of radio industry production in the five years before Pearl Harbor. The miniaturisation and increased efficiency of tube amplifiers enabled consumers to save space while increasing their entertainment options. Moreover, 83 percent of these combinations featured an automatic record changer.[9]

In that statistic lay part of the immediate goad to record innovation. RCA Victor's recreation of the record industry led to competition that shrank profit margins. The unit price of the company's equipment had been cut by more than half between 1936 and 1940, by which time it nearly matched the average of all its licensees.[10] To avoid competition on price alone, RCA Victor needed to offer features that the competition could not match. The main attraction was television, which RCA president David Sarnoff had shepherded to commercial introduction in April 1939. Prewar sets took up considerable space in themselves, however, with the picture tube mounted upright with a mirror to reflect the image to the viewers. It was unrealistic to imagine combining this entertainment with a radio and a 78 record changer in the average living room.[11]

The alternatives for home selection of music were a longplaying record or a changing mechanism for a smaller record. The former was not a new idea at the company, which had researched the technology in 1935 after a disastrous innovation effort in 1931.[12] This format was primarily directed, however, at the classical music audience. These consumers appreciated a mechanical continuity in the longer compositions, but they comprised only a small fraction of the market for records.

Moreover, if a longplaying record was used to compact the multidisc albums of the popular music market, phonographs would lose their advantage over what was offered in radio broadcasts. To maintain the democracy of listener choice embodied in the Victor slogan, 'music you want when you want it', the company's engineers needed to provide an inexpensive and reliable record changer. This challenge had bedevilled them for over a decade. The weight of a stack of 78s on the spindle, and the differently shaped edges of different companies' discs broke down the drive and separator mechanisms. Because of this changers enjoyed the worst technical reputation among all product lines at RCA Victor.[13]

A new record and phonograph would resolve the drawbacks of the 78 and make room in the combinations for television. Given the figures for radio-phonographs, it seemed obvious to Victor executives that now three entertainment technologies would have to fit in one cabinet. These consoles would provide better profit margins, while miniaturising records to increase storage space would also add to sales.[14] It would also provide a record of standard dimensions that would permit more effective interchangeability among the machines and discs of different companies. Because RCA was the only company vertically integrated through record and player production, because it was an industrial leader in both areas and because it had led the drive for standardised equipment in radio, its executives and engineers from Sarnoff on down saw it as the logical innovator of a standardised recorded music system.[15]

To this end it appears that Victor executive Thomas F. Joyce directed the Advanced Development Group in Camden, NJ, to reinvent the record and its player in 1939.[16] By 1941 Ben R. Carson's team had developed a record made of Vinylite and a changing mechanism placed in an oversized spindle. The record was smaller and thinner than the 10-inch 78 used for popular music, enabling the use of the manmade plastic and eliminating the surface noise present in the friction between stylus and mineral filler. RCA engineers and advertising claimed that the changer was the fastest ever, although reviews suggested otherwise.[17] The standard dimensions of the 45, however, assured mechanical reliability at a lower cost if the rest of the record industry adopted it.

This would in turn depend on how optimal the disc was in terms of playing time. To determine that factor, the engineers requested a survey of the records in the company's 'Music America Loves Best' catalog. This included classical pieces divided by the composer, like symphonic movements, and did not include current popular hits. Seventy percent of the classical recordings ran under five minutes, as did 82 percent of the MALB catalog. When all Victor records were considered by sales volume, the figure rose to 96 percent, and it could be assumed that the industry proportion, because of minimal attention to the classical audience, was even higher.[18]

As to improved sound, the issue was moot. Although the lead engineers asserted in 1949 that the goal of higher fidelity had been 'an engineering objective for some time', the cheapness of the player prevented anyone from appreciating the aural qualities of the new record.

If improving delivery of music on record was a relatively objective task of engineering and catalog surveys, defining the parameters preferred by a mass audience was another matter. The question of what constituted fidelity and whether it was a desirable goal arose anew with electrification. Trade demonstrations of a record that more than doubled the frequency range of transcription recordings in late 1931 led to discussions about why the public and many of the attendees preferred more limited response.[19] Tests performed at RCA in 1939 also suggested that the standard would have to be no higher than it was for 78s or amplitude-modulated (AM) radio.

It is not clear whether Victor conducted these tests for its new record project or because the Federal Communications Committee had received 500 applications for FM licenses.[20] High fidelity radio meant that the source of music would have to be improved, while RCA would have to develop and produce transmitters, receivers, and speakers capable of transferring and reproducing that sound. The consequence to RCA as a whole would be a diversion of funds from Sarnoff's commitment to television. If RCA was to invest in FM, therefore, the company needed to see if the public preferred a better quality of sound, and William Perkins surveyed 236 employees at the Camden plant to find out.[21]

The results did not offer an incentive to either improved reception or record innovation. Ninety-four percent of the non-engineers preferred limits under 5,000 hz. Perkins also compared a violinist playing live and through a speaker. Listeners agreed that the open-frequency performance was more realistic, but there 'was unanimous agreement the limited range electrical reproduction [5,000 hz] was more pleasing than the actual performance'.[22]

Although some evidence suggested that these prejudices were learned, not innate, the overall results indicated that technical fidelity was not critical to music appreciation. After the war, however, evidence mounted to the contrary. Decca Records of England began exporting to the United States its 'full-frequency range recording' 78s based on sonar research. By 1947, standard 78 phonograph technology had been developed to the point where RCA's Harry Olson could match a modified home phonograph system to live performances by the Boston Symphony Orchestra.[23] Olson, head of acoustics research at the RCA's Princeton labs, disagreed with the conclusions drawn from the RCA and other tests, pointing out that they were subject to at least six forms of nonlinear distortion.[24]

Olson used live performers playing through acoustic screens. If people preferred a frequency range cut off at 8,000 hz instead of the 20,000 hz understood to represent the outer limits of human hearing, then perhaps 'the public would like an orchestra better if it were surrounded by a light transparent acoustic filter which eliminated the high frequency sound'.[25] That turned out not to be the case, and Olson's 1947 report helped revive research into record systems distortion.[26]

With the product in hand, RCA Victor might have introduced the 45 in 1946. The Home Instrument Division's marketing department pondered ways to sell the system in March 1944, emphasizing issues of cost to win over the public and leaving open the possibility of a longer playing, higher fidelity record later on.[27] By November 1945 the situation had changed. The loss of cost advantages and the potential of a format war with magnetic media chilled prospects for the new record.[28] In the larger context, the year after Japan's surrender featured economic uncertainty, production and employment transitions, and material shortages. And at RCA Victor, when choices had to be made, emphasis would be centered on the television system that Sarnoff had shepherded for fifteen years. At the same time, record sales of discs almost incidentally gave engineers time to improve the fidelity of the 45 – but not the player.[29]

When Columbia introduced its longplaying record for classical music in late June 1948, all accounts indicate the surprise or dismay of RCA executives and engineers. The prospect of competition also raised questions about the best format for home systems. The most damning internal indictment came from Herbert Belar, who had been charged with a wide-ranging record research program in the spring of 1948.[30]

Belar compared the new format to the 78 and Columbia's 33 1/3-rpm LP. Acknowledging ignorance of marketing considerations, he emphasized the company's obligation as industry leader to customers threatened with obsolete record libraries and to new standards of quality, not merchandising. Having calculated that the LP and the 45 would offer sound quality worse than a vinylite microgroove 78, Belar argued unsuccessfully that decisions should not be made on the basis of material conservation or smaller size but in favor of a microgroove plastic record that played at the standard speed.[31]

That was not the issue that irritated RCA president David Sarnoff. Born beyond the Russian Pale, trained in the Talmud, and forced to support his mother and siblings in New York City, he acted on a strong sense of patriarchy in all of his relationships. Paternalism defined his conception of familial, corporate and industrial roles. If the group rallied around the leader who set the standards, all would benefit from ensuing economic success. From this perspective, junior members, in this case William S. Paley and Columbia Broadcasting System, should not confuse the public with competitive formats.[32]

Production of the 45 players began in the fall of 1948.[33] Even then it is possible Sarnoff would have accepted two record formats were it not for criticism of the LPs' sound and Paley's negotiations with NBC's radio star, Jack Benny. On top of the persistent efforts by Paley's minions to obstruct RCA's electronic television, this was the last straw. Sarnoff was 'deeply affronted' by the successful overtures to Benny, and it was his pique over the betrayal of his RCA family that provoked the battle of the speeds, in which RCA Victor would market boxed albums of symphonies for eighteen months in competition with the longplaying records of the rest of the classical industry.[34]

Despite RCA Victor's industrial pre-eminence, vertical integration, and marketing resources, the 45 was rapidly relegated to the childrens' and pop markets. Initially, however, the independent labels in the musics leading to rock and roll were among those most opposed to or least able to adopt the 45 format. Sydney Nathan, owner of King Records, predicted in late 1949 that the 45 would not succeed, for independent record labels had flourished in the 1940s while selling the bulky and brittle 78s.[35] The resistance had less to do with the expense of conversion than with the fact that much of the audience for independent music was least able or least willing to afford new and incompatible record players. Herbert Belar had been right to protect consumers with collections of 78s, but he was thinking of a different class in terms of income and musical taste. Elvis Presley's recordings for Sun Records, as well as many country-western and rhythm'n' blues singles in the first half of the 1950s were released as 78s as well as 45s.[36]

Nonetheless, along with the LP, the 45 did enable more effective diffusion of those musics. The smaller size and lower cost of material, production and shipping were of considerable aid to smaller labels and dealers by the mid-1950s; jukebox manufacturers adapted their machines to the wide-holed 45 in less than two years.[37] Meanwhile the LP dominated not only classical recordings but Broadway soundtracks and the many variants of instrumental and 'mood' musics catering to adults.

Victor intended that the 45 would standardise the record for the great middle-class, 'midcult' market interested in sound better than that from 78s, an audience that would buy popular classical selections as well as popular songs. It would perpetuate a link between high and low musical culture that began with the 78 and continued with the refinement of record changers that could accomodate that format's two diameters. The 45 would also increase the reliability and sound quality of phonograph systems. This in turn would make records more appealing in the competition for consumers' leisure-time.

Victor's intentions came to nought when Columbia went ahead with the LP. As a result the 45 appeared then and remains in the historiography as an embarassed response to the longplaying record, most notable for its contribution to the rise of rock and roll. The perspective offered here suggests a more nuanced view.

Notes

1 R. Gelatt in *The Fabulous Phonograph* (New York, rev. ed. 1965) and O. Read, W. Welch in *From Tin Foil to Stereo: Evolution of the Phonograph* (Indianapolis, second ed., 1976) focus on classical music and the utility of Columbia's longplaying record for that fraction of the record market. A. Millard follows their lead and describes the 45 as 'the driving force of popular music' for the 1950s: *America on Record: A History of Recorded Sound* (Cambridge, New York, 1995), 257.

2 C. M. Burrill, 'Survey of the Phonograph System for Sound Reproduction in the Home: Part I - The Recording System', TR-176, 31 October 1932, and 'Part II - The Reproducing System' TR-183, 25 January 1933, box 116, Accession #2069, Hagley Museum and Library, Wilmington, DE (hereafter Hagley 2069); R. R. Beal to O. S. Schairer, 'Detailed Budget: Research and Advanced Engineering for the year 1935, R.C.A. Manufacturing Company, Inc.', 19 June 1935, from RCA Labs – Bldg. Activities file, Public Affairs Office, David Sarnoff Research Center (DSRC), Princeton, NJ, courtesy Alice Archer. In the Hagley RCA collection I also reviewed boxes #115-121, Technical Reports, 1930-54; #47-50, Engineering Memoranda, 1930-50; and #143-5, »Z« Reports #1-180, 1930s-1960s, all originally from the RCA Camden library.

3 See two articles by F. C. Barton, 'Victrolac Motion Picture Records', *Journal of the Society of Motion Picture Engineers*, 1932, XVIII: 452-5, and 'High Fidelity Lateral-Cut Disk Records', and Discussion, JSMPE, 1934, XX: 179-82.

4 At five dollars a pound, material for a four-ounce record amounted to $1.25: J. Davidson, 'Petrochemical Survey: An Anecdotal Reminiscence', *Chemistry and Industry*, 1956, 19: 395. RCA Victor's Red Seal classical records retailed for two dollars; material accounted for a fraction of the 19 cents attributed to manufacturing costs and overhead. Those expenses were less than a penny for a popular disc: 'Phonograph Records', *Fortune*, 1939, 94.

5 H. Olson, 'Subjective Response-Frequency Tests', report Z-2, p. 1, Hagley 2069, op. cit. (note 2), box 143; 'Music for the Home', *Fortune*, October 1946, 157-9; W. Isom, 'Record Materials', *Journal of the Audio Engineering Society*, 25: 723; 'Loudness of Familiar Sounds: music and speech intensities based upon Bell Laboratories' survey', unnumbered page from *Radio Today*, [1938], in Hagley 2069, op. cit. (note 2), box 290, 'Historical files: Decibels, Spectrums, etc.' This file has since been reorganized in the Historian's series.

6 See H. S. Maraniss, 'A Dog Has Nine Lives: The Story of the Phonograph', *The Annals of American Academy of Political and Social Science* 193 (September 1937: Revival of Depressed Industries), 8-13, and F. Melcher's editorial on the RCA's methods, 'They Built Up Their Outlets', *Publisher's Weekly*, 134, 9 July 1938, 97. Maraniss was an RCA Victor executive.

7 'Comparison Records vs. Instruments, Units, Victor, RCA, and RCA Victor, 1901-1944', Hagley 2069, op. cit. (note 2), Historian's series, box 6, folder 46.

8 'Record Player Production from 1934 to 1942 Incl.', Hagley 2069, op. cit. (note 2), Historian's series, box 9, folder 45.

9 'Automatic Record Changers: Prospect of Post-War Industry Production', 11 July 1944, 1-2, Hagley 2069, op. cit. (note 2), Historian's series, box 9, folder 45.

10 'A Review of RCA Manufacturing Company's Radio Receiver and Phonograph Activities Yearly from 1933 to 1940, inclusive, in Comparison with Sales by All Licensees', 2 April 1941, Hagley 2069, op. cit. (note 2), Historian's series, box 9, folder 49.

11 M. Ritchie, *Please Stand By: A Prehistory of Television* (Woodstock, NY, 1994), 63.

12 Beal to Schairer, op. cit. (note 2), 3; Gelatt, Read and Welch, op. cit. (note 1), 252-4 and 291-3, respectively.

13 'RCA Victor's Overall Automatic Record Changer Activity by Changer Mechanism', 2 October 1945, Historian's series, box 9, folder 45, Hagley 2069, op. cit. (note 2).

14 'Analysis of the Market for Model 'X' Record Changer', 3 March 1944, Hagley 2069, op. cit. (note 2), Historian's series, box 9, folder 45, p. 1.

15 A review of Sarnoff's pronouncements at his Library shows that he regularly justified RCA's existence or its position by its leadership in providing industrial standards. See also B. L. Aldridge to Joe [B. Elliott], 28 November 1945, 2, Hagley 2069, op. cit. (note 2), Historian's series, box 9, folder 45, for comments on the marketing of industrial leadership, and H. I. Reiskind, 'Groove Dimensions of Commercial Records', EM 2409, 10 February 1942, Hagley 2069, op. cit. (note 2), box 48, for the engineering interest in standardisation.

16 Joyce gained credit for the late 1930s revival of the record industry as vice-president for advertising at RCA Victor: R. Sanjek, *American Popular Music and Its Business: The First Four Hundred Years, v. III From 1900 to 1984* (Oxford, New York, 1988), 138; Read and Welch, op. cit. (note 1), 295. As head of the marketing department for Home Instruments during World War II, Joyce solicited the many market analyses prepared by Ben Aldridge; see the above citations from the Hagley Historian's series. It is possible that the 45 innovation was initiated by E. N. Deacon, who was appointed head of New Product Market Research and Analysis in January 1938. See *The Scanner*, February 1938, 3: 30, courtesy of N. Pensiero and since donated to the Hagley Museum and Library.

17 B. Carson, A. Burt, and H. Reiskind, 'A Record Changer and Record of Complementary Design', *RCA Review* 10 (June 1949), 174, 179; *Consumer Reports*, 'New RCA-Victor "45's" – results of CU's first listening tests', May 1949, 197; L. Carduner to the editor and 'Stylus', 'Comment on Comments', *Audio Engineering*, May 1949, 4, 5; P. Goldmark with L. Edson, *Maverick Inventor: My Turbulent Years at CBS* (New York, 1973), 142-3.

18 Carson et al., op cit. (note 17), 183-4.

19 H. Frederick, 'Vertical Sound Records: Recent Fundamental Advances in Mechanical Records on Wax', with Discussion, *JSMPE*, February 1932, 141-63. While the engineers were highly impressed by the realism of the new records, Leopold Stokowski explained how much further re-

produced sound had to go technically and esthetically. In the same issue, see his comments, 'Sound Recording – From the Musician's Point of View', 164-71.

20 T. Lewis, *Empire of the Air: The Men Who Made Radio* (New York: Burlingame, 1991), 278.

21 W. Perkins, 'Phonograph Listening Tests', TR 473, 22 December 1939, Hagley 2069, op. cit. (note 2), box 119.

22 Ibid., 10, 7-8.

23 H. Olson, 'Wide Range Sound Reproducers and the Tanglewood Demonstrations', PEM-79, 18 August 1947, David Sarnoff Research Center Library basement, Princeton, NJ (hereafter DSRCL); also in Hagley 2069, op. cit. (note 2).

24 Olson, op. cit., (note 23), 1-2. The more publicized investigation was by H. Chinn of the Columbia Broadcasting System and psychologist P. Eisenberg, 'Tonal-Range and Sound-Intensity Preferences of Broadcast Listeners', *Proceedings of the I.R.E.* 33 (1945), 9, 571-81.

25 P. Eisenberg, op. cit. (note 24), 2-4; quote, 4.

26 H. Olson, 'Frequency Range Preference for Speech and Music', *Journal of the Acoustic Society of America* (July 1947), 19: 549-55; Carson et al., op. cit., (note 17), 179, credit Olson.

27 'Analysis of the Market...', op. cit., (note 14).

28 B. Aldridge [to J. Elliott], 28 November 1945, Hagley 2069, op. cit. (note 2), Historian series, box 9, folder 45.

29 M. Corrington, 'Tracing Distortion in Phonograph Records', *RCA Review*, 1949, No. 12, 10: 241-53; Carson et al., op. cit. (note 17), 180, 182; 'The Three Types of Phonograph Records – Which to Buy?', *Consumers' Research Bulletin*, November 1949, 18. H. E. Roys notes Corrington's calculations in TR-1006, 18 December 1947, 9, Hagley 2069, op. cit. (note 2), box 121.

30 H. Belar, 'Record Research', lab notebook P-4613, DSRCL, op. cit. (note 23). In it see H. Olson, 'Preliminary Program Phonograph Record Research', witnessed 3 May 1948, and Belar's outline, 'Phonograph Record Research', 1 April 1948.

31 [H. Belar], 'Disc Reproducer Considerations', 9 June 1948, 6, 8, in lab notebook P-4612, 'Disc Record Research', DSRCL, op. cit. (note 23). See also 'Stylus', 'Characteristics of the New 45 RPM Record', *Audio Engineering*, March 1949, 47, for a suggestion similar to Belar's.

32 While the standard biographers and historians of RCA make note of Sarnoff's background and his ego, none of them links these factors explicitly to his sense of personal and corporate destiny. See K. Bilby, *The General: David Sarnoff and the Rise of the Communications Industry* (New York, 1986); C. Dreher, *Sarnoff: An American Success* (New York, 1977); M. Graham, *RCA and the VideoDisk* (Cambridge and New York: Cambridge University Press, 1988), 30-75; Lewis, op. cit. (note 19), 89-118; E. Lyons, *David Sarnoff* (New York, 1966); R. Sobel, *RCA* (New York, 1986).

33 'Historical Files: Firsts: First 45-rpm attachment – production totals', Hagley 2069, op. cit. (note 2), box 291. These files have since been reorganized in the Historian's series.

34 Bilby, op. cit. (note 31), 262-63.

35 See label citations in C. Gillett, *The Sound of the City: The Rise of Rock and Roll* (revised and expanded ed., New York, 1983), 67-96.

36 Sanjek, op. cit. (note 15), 343; 'Executive Opinion: the Influx of Independents', *Business Week*, 15 October 1949, 34-5; 'Elvis: The King of Rock 'N' Roll: The Complete 50's Masters', RCA 66050-2 (1992), in accompanying booklet, [11].

37 C. Gillett, op. cit. (note 35), 101-17; R. E. Mittelstaedt and R. E. Stassen, 'Structural Changes in

the Phonograph Record Industry and Its Channels of Distribution, 1946-1966', *Journal of Macromarketing*, Spring 1994, 31-44; A. G. Bodoh, 'The Jukebox, the Radio and the Record', *Journal of the Audio Engineering Society* (October/November 1977), 25: 837-8.

Andre Millard
Tape Recording and Music Making

Most studies of electromagnetic recording have focused on the technology and its development, but what of its impact on musicians? Historians have interpreted the introduction of magnetic tape recording as a major technological innovation with important social consequences. Tape recording offered many advantages over the existing method of recording on acetate covered discs. It was not only cheaper, it was also much easier to record on tape. This was an accessible technology which permitted more people to enter the professional recording industry. Studies of the origins of rock'n'roll have stressed that this technology empowered the small record studio owners who now had the means to compete with the larger corporate studios on the same technological footing. In this way tape recording was identified as an important facilitator of rock'n'roll music in the fifties – the decade that followed the introduction of magnetic recording after World War II.[1]

Yet owners of recording studios were not the first to buy into this new technology. Among the first purchasers of tape recorders were musicians eager to copy and replay the new music emerging after World War II. Musicians had not been major users of sound recording equipment up to this time, probably because most popular music could be written down. But the jazz, rhythm and blues and rockabilly that was heard in clubs and juke joints in the late 1940s was not written down; it was free form, improvised and constantly changing. The only way to save these performances was to record them.

The existing technology of recording on acetate discs, such as the Presto brand audiodisc system, required expensive machinery and a high level of skill to operate. The input from the microphone was amplified through a vacuum tube amplifier and then cut into the acetate covered revolving disc. The cutting head required constant attention and the user had to decide on several important variables such as the type of cutting stylus, the angle of the cut and the number of threads cut per inch.[2] Although these 'instantaneous' disc recording systems were widely used in commercial applications and in schools, they were far too complex for the amateur. Nor were they suited to operating in dark, crowded conditions that often faced the recorder of popular music.

The prevailing technological system of disc recording mirrored the economic conditions of popular music and the commercial recording industry. Control of the means of producing and distributing musical entertainment was in the hands of a few large, integrated companies. Recording equipment and facilities were monopolized by the six or seven 'major' recording companies. Musicians were not in control of their music; this was the domain of the record producer and the A&R man. The medium of recorded sound was the play-only, fragile 78-rpm shellac disc. The short playing time of this disc basically forced all popular music into a straightjacket of around three minutes.

Magnetic recording offered a release from this restrictive technological system. Recording on steel wire or paper tape covered with iron oxide brought much longer recording times and opened the way for the first economical home recording system since the Edison cylinder recorders of the late nineteenth century. The first practical tape players were made in Nazi Germany and soon made their way into the laboratories, radio stations and record studios of the victorious powers.[3] Within a few years of the end of the war, music aficionados were carrying suitcase sized magnetic wire recorders into night clubs. It was in this way that the most important archival record of the development of be bop jazz was made of Charlie Parker and Dizzy Gillespie in New York jazz clubs.[4]

Wire recorders required a level of skill from the operator and a broken wire could be a dangerous thing, but the recorders were sturdy, portable and produced an adequate quality of playback. Paper or plastic tape was a much easier recording medium to work with, and consequently the first mass produced magnetic recorders that appeared in the late 1940s employed tape. In much the same way that Thomas Edison had little idea what use would be made of his phonograph and who would buy it, the manufacturers of tape recorders only had vague plans for the marketing of this new sound recording system.

Musicians were an early group of customers for the new technology. Initially tape recorders were a tool to learn from, or copy, other musicians' work. Musical education by listening to records was probably as old as the phonograph, but now musicians were free from the monopoly pre-recorded music – the ability to record at home opened up a world of opportunities for them. Sound recording also played an important role in composition, providing an immediate reprise of any new musical idea.

The cheap, portable tape recording therefore became an important factor in the development of new popular music, and the creation of new singing stars, in the 1950s and 1960s. The influential rhythm and blues artist Chuck Berry had an interest in the technical workings of the phonograph and radio, which had provided his musical education. Berry's intimate knowledge of other musical styles was an asset in his development as a song writer. His first big hit for Chess, 'Maybellene,' had strong country overtones – it was originally titled 'Ida Red.' The most important tool in his song writing was a tape recorder which he used to perfect his songs. He first bought a wire recorder in 1951. A $79 reel to reel tape recorder from Radio Shack was his next purchase, and this was used to make the tapes that got him a recording session with Chess records and brought him a string of classic rock'n'roll songs.[5]

Chuck Berry was one of the few rock'n'roll performers who took an active interest in the technology of recording, which was changing rapidly in the 1950s. A few small companies, one of which was called Ampex, had incorporated the new magnetic tape technology in professional studio recorders. These machines were soon in general use in the commercial recording industry, where they stood next to the old disc cutters. Not only could they provide cheaper and longer recordings, but they could also be used to manipulate the recording to change the sound of the music. One popular technique was to double-track the vocal: this involved making two recordings of the same vocal and superimposing the second on the first with a slight delay, giving the vocal more density and depth. This can be heard in Buddy Holly's 'Words of Love' of 1957, which gives the impression of several voices in harmony. Buddy Holly was another musician seriously involved in the technical side of recording – he quickly mastered the studio techniques of making balances and mixing tracks, experimenting with multi-tracking and echoing.[6]

The growing technical awareness of the artist was an important factor in moving the balance of power from the engineers and record producers employed by the record company to the performers themselves. The selection of material to be recorded was no longer the unquestioned preserve of the A&R man. In the first part of the twentieth century, the song was everything, a property to be recorded by many artists and the source of profits through royalties. Sheet music was copyrighted and protected. The recording was now the important thing – the property and the profit maker. In the 1950s an artist like Chuck Berry or Buddy Holly would come to the studio with his own material and his own ideas about playing it. The song was his property. He might have to split the authorship with the owner of the record company or with the deejay who promoted his music, but he was still the source of the material he recorded. And the format of the song had changed; no longer on a sheet of music, it was now contained on a reel of magnetic tape.

During the 1960s the true potential of tape recording was discovered in recording studios on both sides of the Atlantic: Capitol Studios in Los Angeles and Abbey Road studios in London were two hot beds of activity as engineers and artists collaborated in the development of multi-tracking and tape editing. The Beach Boys' 'Petsounds' and the Beatles' 'Sgt Pepper's Lonely Hearts Club Band' are the best known products of this artistic and technological revolution in recording popular songs.[7] At the same time, the growing availability of tape recorders encouraged performers to make their own recordings at home.

The band that backed Bob Dylan during his infamous 1966 tour, which featured his first electrified performances, returned to the United States in 1967 and rented a house in upstate New York. From March to December they recorded in the basement of the house on a two-track reel to reel tape recorder. The house was known as The Big Pink, the group was later blandly known as the Band, and the recordings were released as 'Music from the Big Pink,' a landmark 'back to basics' recording which can compare with any of the classic albums of the 1960s.[8]

The technology of tape recording was progressing rapidly at the end of the 1960s, along with the ambitions of technically minded musicians. While Chuck Berry and Buddy Holly were putting tape recorders in their homes, bands like the Beatles and Beach Boys had complete home recording studios that were the equal of any record company.

The musical highlights of the 1960s owed a great deal to the flexibility of tape recording and the growing accomplishments of musicians as recording engineers. The control rooms of the big studios were no longer out of bounds for musicians and the home studios of musicians were approaching the technical capabilities of the corporate studios. Yet the final product of all this work was still a revolving disc because whatever the origin of the master tapes they were still only one part of a process that was still firmly in the hands of the record companies. Home studios gave artistic control into the hands of musicians but production and marketing were still the preserve of corporations. The sound of the music might be different but it was still made and sold in the same old way, a tradition which kept popular music in the same old corporate nexus of independent companies and media conglomerates.

This hegemony was to be overturned by an innovation in tape recording that brought forward the most serious disruption of popular music and its business since Bill Haley and Elvis Presley upturned the music world in the early 1950s.

The introduction of the tape cassette marked the beginning of a new era in both machinery and music. This technology began life as a format for the dictating machines used by businessmen. The cassette was much easier to handle than the clumsy reels of tape and had the added advantage of easy portability. Several European concerns employed the tape cassette in small, personal dictating machines. In 1962 the Philips Company developed a cassette in which the tape was half as wide as the standard 1/4-inch tape and ran between two reels in a small plastic case. The tape moved at 1 7/8 inches per second, compared with the 3 3/4 on eight-tracks and home tape recorders. Its very slow speed put more words on the tape but paid the price in limited fidelity.

The Philips compact cassette was introduced in 1963. During the first year on the market, only 9,000 units were sold. Philips did not protect its cassette as a proprietary technology but encouraged other companies to license its use. The company required that all users of its compact cassette adhere to its standards, which guaranteed that all cassettes would be compatible. An alliance with several Japanese manufacturers ensured that when the format was introduced for home use in the mid-1960s, there were several cassette players available. The first sold in the United States were made by Panasonic and Norelco. They were quickly joined by other companies as the compact cassette took hold. By 1968 around eighty-five different manufacturers had sold over 2.4 million cassette players worldwide. In that year the cassette business was worth about $150 million.[9] The Philips compact cassette became the standard format for tape recording by the end of the decade and the cassette players – from Sony Walkman personal stereo to the ubiquitous 'boom box' – became the sound machine of choice in the 1980s.

Its name gives us a clue to its wide appeal. It is small enough to fit into a shirt pocket, yet it can still hold 45 minutes of music on each side of the tape. It is very easy to use – one has only to insert it into the player, usually by lifting up a lid, and press the PLAY button, the mechanism then engages, and the tape spools turn.

The major drawback of the audio cassette was that it was a low-fidelity medium. It had a poor frequency response when compared with the disc, and its playback was marred by a loud and annoying tape hiss. The serious home listener could not be persuaded to accept it with these disadvantages, and the cassette was seen as a means of bringing portable sound to the less discriminating user – a tape version of the transistor radio. Yet over time there was a steady improvement in all parts of the cassette recording system.

Cassette tape was dramatically improved in the decade after its introduction in 1963. The chemical companies developed magnetic compounds that could hold more information, gradually perfecting the basic ferric oxide compounds with the addition of cobalt mixtures and then introducing chromium dioxide as the magnetic material on the tape. This provided superior recordings, and more information on the tape permitted the recording of stereo signals. A lot of the annoying surface clutter of cassette recordings were effectively eliminated by the noise reduction systems like Dolby.[10]

Japanese manufacturers led the way in incorporating the cassette into home stereos. The transistor and the cassette became complementary technologies; the cheap radio/phonograph combination soon came with a cassette player built in, which could record any output of the unit. By the 1970s the cassette player was incorporated into the high-fidelity equipment purchased by the more demanding home user. This was also the work of the Japanese manufacturers, who were led by Sony, Matsushita, and Nakamichi in producing cassette tape decks which could compete with the reel-to-reel units in sound quality.

By the 1970s the high end cassette recorders were approaching the standards set by reel to reel recorders. In the world of home recording, the cassette had become good enough to record commercially and cheap enough to become an attractive alternative to the top of the line home recorders. To the musician the portability and lower price of the tape cassette more than made up for any deficiencies in fidelity. Tape recording had by definition been reel to reel up to this time, but the cassette was to change that.

Reels of tape were the medium of studio recordings but they were too cumbersome to be used in the marketing of music: this was the preserve of the 7-inch acetate disc, which was hawked by musicians and promoters from agent's office to record studio to radio station. The all important demonstration song – the demo – was the currency of commercial music, providing the material for the transactions of the music business. The popularity of the cassette in home taping was responsible for it becoming the recording medium of choice for musicians. And soon, instead of sliding a sheet of music or a 45-rpm disc across a desk, the promoter could now hand over a tape cassette. Like the 7-inch disc before it, the cassette was an accessible recording medium for the basic unit of popular song, but it had the advantages of convenience and low cost and it was already in use in everyone's home.

In this way the cassette became the demo of the 1980s and 1990s, the tool with which musicians and song writers traded their wares. Ease and cheapness of cassette duplication meant that every working musician could have numerous copies of his work at hand, to give to record producer or club owner, or even to sell to the fans. For those bands who had already used cassettes for their demos, it was a very short step to actually making their own commercial recordings. For many new wave bands of the 1980s the cassette was the only means to market their music because record companies and radio stations would not accept them. The economics of mass producing music cassettes was favorable to the musician, and in many cases a band could make more money selling their own cassettes on the road than through a record contract with a large company.[11]

For all these reasons the cassette became the means of introducing a new and much more radical popular music. Rap is now on record and compact disc, but it could only have begun on cassette tape. As an underground music it relied on home taping and duplicating facilities of the musicians themselves. Rap emerged from the black community and stood alongside such ghetto arts as graffiti writing and break-dancing. It came from the urban areas which had a strong cassette culture: boom boxes became associated with black youth in the 1980s, especially after African American film directors made them a symbol of urban life in films like 'Do the Right Thing' by Spike Lee.

Although it could be argued that rap came at the end of a long road of underground black music, which started with the field hollers of the slaves, it was nevertheless heavily influenced by the music technology of the 1970s. It began in dance halls and clubs in New York and Brooklyn, where disc jockeys played records and talked to the crowd over the songs. They employed the common disco setup of two turntables and a mixer: one turntable could be used for the rhythm and the other for the lead vocal of instrumental.[12] Although disco and rap were so firmly based on the revolving disc as to be defined by it, the work of the creative disc jockeys was saved on the ubiquitous cassette tape.

The first commercial rap singles appeared at the end of the 1970s, notably 'Rapper's Delight' by a group called the Sugar Hill Gang on their own independent Sugar Hill label in 1979. Dismissed by the major record companies as a fad and 'too black' by radio stations, rap could only exist underground, and the only format in which underground music could exist in the United States was the cassette tape.

Rap was seen by its creators as subversive, overthrowing the musical status quo and undermining the monopoly powers of the large record companies. The pioneer rappers were more concerned with duplicating the raw spontaneity of dance parties than making commercial dance records. The rapper Fab 5 Freddy criticized many of the commercial rap recordings: 'People weren't capturing the essence of hip-hop that was at the parties with the scratching and all that.'[13] Rap and hip-hop was underground music in that it was produced locally and not immediately co-opted by the media conglomerates and marketed on records. Although the record was the source of the music, the music was not put on records because that was the preserve of corporations with their recording studios and stamping presses.

The format of rap and hip-hop was the cassette tape. It was easy to record, and everybody knew how to do it. The wide availability of double cassette decks made it even easier to re-record and copy songs. The cassette was the ideal vehicle for a non-commercial music made by teenagers. Hardly any rapper had access to a recording studio, but everybody had a boom box, and the universal practice of home taping had diffused the essential skills of recording. Rap was similar to new wave or punk music which had also emerged from disgruntled urban youth in the 1970s. Rap had the distinctive low tech sound of alternative music, although this was more the result of lack of money and sophisticated equipment than the conscious decision to subvert the technology of the studio.

Once the rap song had been recorded on a cheap, transistorized cassette deck, there remained the old problem of distributing it. Rap was ignored by black stations, who stayed in the more predictable arena of gospel, soul, and funk. It was also avoided by all the major record companies. The rappers duplicated tapes on dual cassette recorders and marketed their own songs in the urban ghettos of the Northeast. Instead of nationwide distribution networks, rap musicians and their friends and relatives hawked their tapes in the neighborhoods.

Despite the lack of radio play and the primitive marketing system, rap recordings slowly attracted public attention. Spread by word of mouth and the sale or exchange of cassettes, rap's popularity gained momentum in the early 1980s. Rap recordings were made in apartments and in houses in New York and the Bronx. A rap record company was any organization which had a room to record in and one of the latest inexpensive, 'professional' cassette tape recorders with equalization controls and a couple of microphone inputs.

During the 1970s both reel-to-reel and cassette recorders incorporated technologies developed in the professional recording studios in the 1960s. One could record on four tracks and copy from track to track in the same way that recording engineers worked in the high-tech studio. The top of the line cassette models came with all the controls found on professional machines: VU and peak meters, several forms of noise-reduction systems, and various filters. Some units came with their own mixing consoles built in. The home listener could buy condenser microphones and monitoring loudspeakers exactly like the ones used in professional studios. In the nineteenth century inventors had hoped that the phonograph would enable the Victorian gentleman to record the music made in his home, but now it was possible to turn a home into a recording studio.

What made rap so significant in the history of recorded sound was that technology had completely taken over the process of making the music instead of just recording and reproducing it. As the authentic folk music of the black masses, rap could not be the product of musical instruments in expensive high-tech recording studios. The techniques of recording had seeped down to the street corner and home studio via the ubiquitous tape cassette. Rappers made their music with the equipment at hand: the home stereo. They appropriated the technology of turntable, microphone, and cassette deck and used it in ways that the manufacturers had never intended.

The cassette tape recorder played a major part in the creation and dissemination of rap in the 1980s and of industrial and house music of the 1990s. Today popular music is not only recorded on tape – it is created out of pieces of tape. The new art form of sampling music is a direct consequence of the growing bond between musicians and tape recording.

The convenient cassette tape that spread the rap sound from city to city in the United States has now made it a global music, heard in Europe, Africa, the Far East, and South America. Calling the cassette the transistor radio of tape recording rang true; the cassette player was a vastly successful commercial product that was manufactured globally. The ease of entry into cassette technology ensured that cassettes were made and sold in every country, and thus the cassette became a world standard for tape recording. No recording medium was, and still is, so universally used as the cassette. In fact the two great consumer constants that can be found all over the globe are the tee shirt and the cassette tape.

The cassette in the hands of musicians and entrepreneurs was to have an even more disruptive affect on the music industries of the Third World than it had in the United States. The first cassette recorders heard in the vast subcontinent of India were brought home by emigrant workers from the Persian Gulf. They were the basic boom box configuration of radio receiver, cassette recorder, amplifier and twin speakers. These machines started a revolution in Indian music as the cassette quickly supplanted the vinyl or shellac disc. The latter were products of the multinational record companies like EMI, who dominated Indian recorded music with the imperial ease of an advanced, technologically rich organization. The technological system of recording on acetate master discs favored the company who alone had the resources to build studios. In this way a company like EMI could dominate Indian popular music from its corporate headquarters in London.

The ease of recording and duplicating cassettes allowed small companies to enter the industry and break up the hegemony of popular music on disc. The compact cassette enriched the regional music of India by destroying the monopoly of the large record companies, enabling new voices to be heard. Mass duplication of cassettes and widespread piracy of recorded sound products restructured the Indian entertainment industry and brought significant changes to popular music. In his book on this transformation, Peter Manuel coined the phrase 'cassette culture' to describe the economic and social consequences of the new form of tape recording, and to show how it helped create new musical forms in India.[14]

The universality of cassette tapes has made it the ideal medium for the diffusion of musical styles through recordings. All the world uses the same standard compact cassette and thus a recording can be taken and played anywhere. In tune with the changing technological times, diffusion with cassette tapes is formidably faster and much more diverse than the technology it replaced. The revolving disc was the vehicle to spread jazz and rock'n'roll far from the borders of the United States across the Atlantic to the industrial West. The cassette has accomplished the creation of a world music through its global availability.

Africa is another example of the influence of the cassette on popular music. Although disc recordings had been made available in the continent, African music was rarely commercially recorded on disc for the benefit of the rest of the world. While Africans could listen to European and American popular music, the distinct sounds of the South African townships or West African cities were not available outside the continent. Musical influences came into Africa – notably the discs of Cuban rhumba music which played such an important part in the development of Zairian *soukous* music – but rarely out of it.[15]

The cassette culture provided the means for African music to enrich the jaded popular music of the West in the 1970s. African percussion, always a favorite of record producers, began the movement. This can be heard in the landmark Joni Mitchell album 'The Hissing of Summer Lawns' (1975). Peter Gabriel and Paul Simon provided the vanguard for the incorporation of African rhythms into popular music in the 1980s. Paul Simon's album 'Graceland' (1986) was the single most important facilitator of diffusion of African music. The liner notes for this album tell us that his interest in African music was aroused when a friend gave him a cassette of 'Gumboots: Accordion Jive Hits.' in 1984. 'Graceland' was such a commercial success that it opened the way for several African artists to enter the mainstream of Western pop.

In the 1990s we now talk of 'world music,' in which influences and artists from all over the world contribute to commercial music in Europe and the United States. Performers from the Third World have already experienced the pinnacle of success in the competitive music business of the West – the Algerian singer Cheb Khaled had an European hit with 'Didi.' The humble compact cassette is the driving force for changes in the sound of popular music in the last two decades. While it is being gradually overtaken by the compact disc as the medium of choice in the West, its continued presence in Africa, Asia and the Middle East will ensure that it remains the main vehicle for diffusion of music for the next two decades.

Notes

1 Hugh Mooney, 'Just Before Rock, Pop Music 1950-1953 Reconsidered,' *Popular Music & Society*, 1974, 3: 93-94; C.A. Schicke, *Revolution in Sound: A Biography of the Recording Industry* (Boston, 1974), 114-115.

2 For example see *How to Make Good Recordings* (New York, 1940).

3 For an overview of the history of tape recording see A. Millard, *America on Record* (New York, 1995), chp. 10.

4 Dean Benedetti started with a portable acetate disc cutter and graduated to a reel-to-reel

recorder in 1948. Listen to 'The Complete Dean Benedetti Recordings of Charlie Parker,' *Mosaic Records*, 1991.

5 C. Berry, *The Autobiography* (New York, 1987), 103, 107.

6 J. Goldrosen, *The Buddy Holly Story* (New York, 1979), 68.

7 The best account of the Beatles recording sessions is M. Lewisohn, *The Beatles Recording Sessions* (New York, 1988).

8 L. Helm, S. Davis, *This Wheel's on Fire* (New York, 1993), 151.

9 *Billboard*, 6 Nov. 1982, wc3.

10 R. Dolby, 'An Audio Noise Reduction System,' *Journal of the Audio Engineering Society*, Oct. 1967, 15: 383-388.

11 Beam Collection of oral histories of musicians, University of Alabama at Birmingham.

12 For an introduction into the origins of this music, see H. Ellis and M. A. Gonzales, *Bring the Noise: A Guide to Rap Music and Hip Hop Culture* (New York, 1991).

13 J. Morley, 'Introduction,' in L.A. Stanley (ed.), *Rap: The Lyrics* (New York, 1992), XVI.

14 P. Manuel, *Cassette Culture: Popular Music and Technology in North India* (Chicago, 1993), 89.

15 In the 1950s several South African tunes were popularized in the West, and the penny whistle became a commercial instrument in pop music. For a good overview of the development of African popular music consult G. Stewart, *Breakout: Profiles in African Rhythm* (Chicago, 1992).

James P. Kraft

MUSICIANS AND THE SOUND REVOLUTION: BUSINESS, LABOR, AND TECHNOLOGY IN AMERICA, 1890-1950

Scholars from various fields and professions have studied the history of musicians, yet few have placed musicians within the context of the various technological and organizational changes that transformed the leisure business in the twentieth century. For the past decade I have made my own attempts to understand how invention, innovation, and expanding corporate power in the realm of entertainment intersected with the lives and work experiences of American musicians.[1] In this essay, I have tried to summarize my work on this subject and identify its theoretical framework. In doing so, I hope to encourage new studies that explore the impact of technological change on work, culture, and society.

Stage to Studio: Musicians and the Sound Revolution, 1890-1950 (Johns Hopkins, 1996) begins in the late nineteenth and early twentieth centuries, in an era of ragtime music, silent films, and vaudeville theater. In many ways, this was a 'golden age' for musicians. In cities and towns across America, the demand for musical services often exceeded the supply of skilled instrumentalists. In contrast to skilled workers in many other trades and professions, American musicians benefited from working for small businesses whose successes or failures depended on the performances of their employees. The main threat facing instrumentalists at the turn of the century was not large corporations with labor-saving technology and strong employer associations, but their own reluctance to recognize and act on their common concerns as workers. Once they did realize their common problems, they built a union – the American Federation of Musicians (AFM) – strong enough to protect their rights and interests. I compared this environment to that of 1925 to 1950, and in doing so uncovered a story of great change. The introduction of sound movies, network radio, and high-fidelity recordings transformed the musicians' world, turning a diffused, labor-intensive, artisanal structure into a centralized, capital-intensive, highly mechanized one. Technologically driven developments in the entertainment business affected wages, working conditions, patterns of hiring, definitions of skills, and above all job opportunities. They brought higher incomes and improved standards of living to many, and fortune and fame to a few, but for the majority technological developments meant dislocation and restricted or lost opportunity.

The coming of sound movies left the biggest imprint on the history of musicians. In 1927, approximately 25,000 musicians – roughly a quarter of all professional instrumentalists in America – worked in theater orchestras, where they enlivened silent films and vaudeville acts. In one fell swoop, and with little regard to seniority of skill levels, sound movies eliminated thousands of theater jobs. By the mid-1930s, opportunities in theaters had all but vanished. The fact that innovation created a few hundred new jobs in Hollywood film studios was small compensation for the elimination of a major avenue of musical employment.

The rise of the radio and recording industries also had major implications for the work environment and market power of musicians. Radio initially provided musicians many new employment opportunities. In the 1920s stations across the nation employed local bands and orchestras to fill airtime and help advertise consumer products. This situation began to change with the rise of national radio networks and the introduction of new recording processes, both of which gave broadcasters alternatives to 'live' talent. Throughout the 1940s small and medium-size stations replaced local staff orchestras with big-city network programs and long-playing, high-fidelity records. In the 1940s and 1950s competition from television and the development of tape recording presented new incentives for displacing musicians. By 1957 the number of musicians with full-time employment in radio had fallen from roughly 2,500 to less than 500.

Technological and industrial change, of course, does not singlehandedly explain the musicians' story. The role of government proved crucial in shaping the impact of invention and innovation. In the 1940s, government policy evolved in ways that eroded the AFM's bargaining power and thus undermined the musicians' ability to protect their wages and working conditions. The Lea and Taft-Hartley Acts proved particularly detrimental to the status of musicians. Passed in the wake of one of the largest strike waves in American history, these acts outlawed several AFM practices, including the practice of demanding 'minimum-size' orchestras in radio. Labor legislation of the 1940s, I have argued, not only reflected concerns about industrial relations; it also showed that lawmakers had interests and ideological perspectives that contrasted sharply with those of workers and their union leaders.

Disaffected musicians did not simply watch these developments capsize their lives. Mainly through their union, musicians sought to control the forces of change. For a decade after the introduction of sound movies, change was so rapid and overpowering that union leaders only engaged in rearguard actions. But once a sense of stability settled over the 'music sector' of the economy, they aggressively resisted management control of new production technologies. In the 1940s, under the leadership of James C. Petrillo, the union won major concessions from industry and in the process pioneered new patterns in labor relations. By midcentury, however, in the face of new technological changes and a wave of antiunionism, the musicians' efforts to protect jobs and income had suffered major setbacks and they and their union were in retreat.

Simply put, then, my work addresses two haunting questions in labor history: What impact did technological change – especially the kind of change that increased worker efficiency and productivity and thus benefited employers and consumers – have on workers, and how successfully did workers cope with that impact? In addition, it asks about the actual as well as the proper relationship among government, business, and labor. Should the state be relatively neutral in matters affecting business, labor, and consumers, or should it intervene in those matters in behalf of one or another of the interested parties? And if it should intervene, when and in whose behalf should it do so?

Answers to these questions have proved elusive. In some industries, technological change has simplified work tasks to the detriment of workers, while in others it has created demands for new skills and increased the challenge of work as well as labor's bargaining power. In all of these industries, workers struggled, with uneven success, to turn technological history to their own advantage. The role of the state in industrial relations, of course, has never been static. In some instances, state intervention has been to the advantage of workers, but in others it has left them powerless and demoralized.

The history of musicians sheds lights on all these scenarios. It shows the ambiguous nature of the changes that invention and innovation have produced, and the varying ways that workers have responded to changing business conditions. It suggests, too, that workers and their unions have generally accepted innovation as inevitable even as they struggled to mitigate its most negative implications. As for the role of government in the history of musicians, it cut two ways. In the 1920s and 1930s, federal legislation protected the interests of instrumentalists by restricting the use of recorded music on radio. In the 1940s, however, new legislation eroded the ability of musicians to control the impact of new technology on their employment.

In terms of a theoretical framework, my work draws on different perspectives and insights. I found much explanatory power in the writing of Alfred D. Chandler, Jr., still the foremost figure in the field of business history. Chandler's most notable work, *The Visible Hand: The Managerial Revolution in American Business*, traced a major shift in the American economy during the ninteenth and twentieth centuries. Beginning with the rise of the American railroads, Chandler explained, an economy marked by numerous small competitive firms began to give way to an oligopolistic framework driven by large capital-intensive, vertically-integrated firms with multidivisional, decentralized administrative structures. In the new environment, production became more centralized and markets more closed.[2]

The 'music' sector of the American economy clearly evolved along these lines. By the mid-1930s, the film industry was dominated by a handful of major motion picture studios with interests in the production, distribution, and exhibition sectors of the film industry. The radio and recording industries had also become oligopolistic. In the 1940s, as entrepreneurs in the leisure business expanded their interests, the lines separating film, radio, and recording grew harder to see. The centralized and integrated nature of the entertainment business partly accounted for the high level of cooperation between industry leaders, and thus helps explain the growing problems of musicians. In short, the Chandlerian perspective brought a much needed sense of structure and theory to the world of musicians. It was not just technology that drove the history of musicians, but the strategies of entrepreneurs and the evolution of business firms.

Unfortunately, Chandler had little to say about the relationship between business development and American workers. Chandler's neglect of labor has no doubt suggested to some readers that industrial change was mostly beneficial and fair, the way things ought to be. But even a glance at American labor history reveals that thousands of industrial workers were unhappy with the course of capitalist development. Indeed, from the steel mills of Homestead, Pennsylvania, to the silver mines at Coeur d'Alene, Idaho, scores of workers forfeited their lives resisting the new industrial order. To understand what industrial change meant for American musicians, and how musicians responded to change, I had to look to other explanatory paradigms.

The Marxist perspective proved particularly insightful. In volume one of *Capital* Marx argued that capitalist development not only centralized industrial production and closed markets but subordinated the interests of workers to those of management. Marx explained, all so eloquently, that capital's use of new technology tended to simplify work tasks and thereby reduce skill levels to the detriment of workers. He showed, too, that mechanization generated labor conflict and a sense of social alienation. Marx also challenged notions about the role of the state in workers' lives. Capital manipulated state power, Marx explained, in ways that undermined the collective strength of labor groups. In several ways, then, the Marxist perspective helped me raise questions about the impact of sound technology on work and labor relations and helped to find new meaning in the politics of the 1930s and '40s.[3]

Works that refined and extended the Marxist view of technological history also contributed to the intellectual scaffolding. Harry Braverman's *Labor and Monopoly Capital: The Degradation of Work in the Twentieth Century* was particularly influential. Braverman explained that new technology tended to separate the 'conception' of work from its actual 'execution.' If this process did not deskill workers, Braverman suggested, it divided workers in ways that undermined labor solidarity. Other scholars, especially Michael Burawoy and Richard Edwards, helped me see that technological change could create new and highly skilled jobs that paid exceedingly well but offered little in the way of job satisfaction and security. David Noble convinced me that technological change could destroy as many jobs as it created, and Dan Clawson helped me see that impersonal bureaucracies no less than mechanization increased management's control over work environments.[4]

Unfortunately, these studies of the work environment treated workers in rather impersonal terms. To bring a more human dimension to my study of musicians as laborers, I borrowed lessons and themes from the 'new' labor history. This school of thought, which emerged in the 1960s and 1970s, nearly obliterated traditional approaches to labor history. Led by pioneering figures like David Brody and David Montgomery, the 'new' labor historians put workers rather than unions at the centerstage of history. To trace the experiences of labor groups, they focused on workers' day-to-day habits and the culture of the workplace. They focused, too, on the diverse ways that workers responded to change. The 'new' labor historians also taught me the value of nontraditional sources – oral history, diaries, and the labor press – as investigative tools.[5]

I tried to combine these historical approaches with the 'contextualist' view of technological history. Unlike 'internalists,' who focused on the internal design of machinery, contextualists have looked at how the development of technical artifacts intersected with larger patterns and processes in history. Contextualists explored relationships between technology and society. In doing so, they rejected the notion that new machinery embodied progress and emphasized that technological change tends to protect the power and well-being of some groups at the expense of others. This school of historical writing is often traced to Lewis Mumford's *Technics and Civilization*, published in the 1930s. In this and other works, Mumford argued that invention and innovation reflected social, cultural, and economic structures and processes. He made it clear that technology is a value-laden thing, a socially constructed force.[6]

The contextualist approach helped me see that mechanization in the field of entertainment embodied the views and interests of employers and consumers much more than musicians, but it also suggested that musicians could turn technology to their own advantage. The AFM's ability to regulate remote broadcasting in the 1940s offers a good example of how musicians gained a measure of control over the course and impact of technological history.[7]

Finally, there were lessons to be learned from the post-structuralists who rely on discourse analysis to understand history. Drawing from the work of Michel Foucault and Jacques Derrida, post-structuralists have argued that language is less a collection of words with definite meanings and more a system of signs or symbols that people manipulate, consciously and unconsciously, to affix meaning to their lives. In other words, they suggested that language *shapes* rather than simply *reflects* social reality. The 'linguistic turn' in historical writing helped me see that musicians' lives were bound up in language as well as technological and organizational change. Throughout my work, I have tried to analyze rhetoric in press statements and trade papers in ways that illuminate the different perspectives and interests of musicians and their employers. These differences in outlooks and value systems, I explained, precluded mutually beneficial compromises at the bargaining table.[8]

In sum, my work draws on several lines of analysis to explore the impact of technological and industrial change. It demonstrates, quite conclusively I think, that the benefits of technological advances in the entertainment business were spread unevenly, according to power relations among the groups most affected by such advances. I have therefore ended my work on a cautionary note. Is technological change liberating or enthralling? The effects of technology on work and workers have varied and will no doubt continue to vary. But the history of musicians certainly challenges the uncritical assumption that new technology automatically means social and material advancement, or more satisfying work.

Notes

1 J. P. Kraft, *Stage to Studio: Musicians and the Sound Revolution, 1890-1950* (Baltimore, 1996).

2 On topics and themes in the history of business enterprise, see A. D. Chandler, Jr., *Strategy and Structure: Chapters in the History of the American Industrial Enterprise* (Cambridge Mass., 1962); *The Visible Hand: The Managerial Revolution in American Business* (Cambridge Mass., 1977); *Scale and Scope: The Dynamics of Industrial Capitalism* (Cambridge Mass., 1992); T. C. Cochran, *Business in American Life: A History* (New York, 1972); and M. Blackford, A. Kerr, *Business Enterprise in American History* (Boston, 1994).

3 K. Marx, *Capital* (New York, 1977), vol. 1, pts. 3 and 4.

4 H. Braverman, *Labor and Monopoly Capital: The Degradation of Work in the Twentieth Century* (New York, 1974). See also D. Noble, *Forces of Production: A Social History of Industrial Automation* (New York, 1984); R. Edwards, *Contested Terrain: The Transformation of the Workplace in the Twentieth Century* (New York, 1979); M. Burawoy, *The Politics of Production* (New York, 1985); D. Clawson, *Bureaucracy and the Labor Process: The Trans formation of U.S. Industry, 1860-1920* (New York, 1980).

5 On approaches to labor history, see D. Brody, 'The Old Labor History and the New: In Search of an American Working Class,' *Labor History*, 1979, Summer 111-26; D. Montgomery, 'To Study the People: The American Working Class,' *Labor History*, 1980, 21: 485-512; and L. Fink, ' "Intellectuals" versus "Workers": Academic Requirements and the Creation of Labor History,' *American Historical Review*, 1991, 96: 395-421.

6 L. Mumford, *Technics and Civilization* (New York, 1934). On approaches to the history of technology, see S. H. Cutcliffe, R. C. Post, *In Context: History and the History of Technology: Essays in Honor of Melvin Kranzberg* (Bethlehem Pa., 1989); T. Hughes, *American Genesis: A Century of Invention and Technological Enthusiasm* (New York, 1989); and M. R. Smith, L. Marx, *Does Technology Drive History? The Dilemma of Technological Determinism* (Cambridge Mass., 1984); C. W. Pursell, *The Machine in America: A Social History of Technology* (Baltimore, 1995); R. Schwartz Cowan, *A Social History of American Technology* (New York, 1997).

7 In the 1930s, local unions in many cities demanded that if radio stations broadcast musical performances from hotels, theaters, and dances, proprietors of those places paid performing musicians' supplementary wages. In the 1940s, the AFM periodically prohibited remote broadcasts to protect the wages and job opportunities of musicians in network-affiliated stations.

8 See L. R. Berlanstein, *Rethinking Labor History: Essays on Discourse and Class Analysis* (Urbana, 1993).

Mark Katz

AESTHETICS OUT OF EXIGENCY: VIOLIN VIBRATO AND THE PHONOGRAPH

Violinists have been using vibrato for centuries. But for more than three hundred years this gentle, shimmering effect was treated as an embellishment, considered artistic only in its subtle and sparing use. In the first decades of the twentieth century, however, the practice changed. Professional, classically-trained violinists began using vibrato more conspicuously and nearly continuously, transforming a special effect into an integral element of violin sound.

This dramatic change is something of a mystery. Although many have commented on the matter, few have suggested answers, and none has explored the issue in any depth. In this essay I will examine a pervasive, but rarely investigated, influence in modern music making, and in doing so offer an explanation for this shift in performance practice.

But first, we should consider the rise of what I call the 'new' vibrato. Before the twentieth century, it was the bow, not the vibrato – in other words, the right hand, not the left hand – that violinists considered the most important means for musical expression. Indeed, the bow was often called the 'soul' of the violin.[1] Treatises rarely mentioned vibrato, and did so usually only to circumscribe its use. Typical in this regard is an 1832 'Violinschule', by the German violinist and composer Louis Spohr, which instructs that the pitch fluctuation in using vibrato 'should hardly be perceptible to the ear'.[2] Some treatises provided musical examples in which the recommended use of vibrato was notated (usually with wavy lines under the notes). An exercise in Pierre Baillot's 1834 'L'art du violon' prescribes vibrato for only nine notes among dozens in the 15-measure example.[3] Charles de Beriòt's example in his 1858 'Méthode de violon' is even more frugal, suggesting vibrato on only a very few of the highest notes in the excerpt.[4] In following decades the eminent violinists Ferdinand David and Joseph Joachim continued the cautionary approach towards vibrato: David wrote in 1864 that 'one must guard against [its] too frequent and unmotivated use', while in 1905 Joachim admonished readers that 'a violinist whose taste is refined and healthy will always recognize the steady tone as the ruling one'.[5] Even the first tutor devoted solely to violin vibrato – written in 1900 by the English pedagogue Archibald Saunders – concurred with the prevailing view of vibrato's limited role. Saunders advises that the violinist 'should avoid its use altogether in rapid runs [and] bear in mind that good violin tone is possible without the employment of this fascinating embellishment'.[6] We can see, then, that before the twentieth century violin vibrato was deemed entirely optional, effective only when applied to selected, usually longer, notes in a melodic line, and tasteful only when its pitch extent (or width) was kept to a minimum.[7]

The earliest indications of a general shift towards a more prominent vibrato come from 1908, as reported by two British writers. Both regarded the trend ambivalently. One called the vibrato a 'lovely adjunct to the violinist's equipment', but objected to its overuse, and observed that the hands of too many violinists trembled 'like jelly on a plate of a nervous waiter'.[8] The other could muster only enough enthusiasm to suggest that, 'when we consider that the world's greatest violinists all use it more or less, it surely must have some virtue'.[9]

As vibrato became more prevalent through the 1910s, some began to argue that it was no longer an option. An influential early work on the subject – Siegfried Eberhardt's 'Der beseelte Violinton' from 1910 – argues that 'artistic finish in playing is impossible without a correctly made vibrato'.[10] An American manual similarly claimed in 1917 that 'no matter how fine a tone one may draw, if it is not vitalized, so to speak, with a gentle and refined vibrato, the tone produced will be dull and expressionless'.[11] Nevertheless, many were still reluctant to sanction an intense, continuous vibrato. For example, a 1916 article from the British string journal *Strad* warned that 'an enthusiastic and passionate vibrato at uneventful moments is as senseless as recitation of the alphabet with intense emotion would be, and equally nauseous'.[12] While this last remark seems simply to appeal to common sense, it nevertheless reveals the belief that vibrato is properly an ornament, an attitude less and less reflected in the practice of the day.

While critics and teachers continued to mount objections to the new vibrato, fewer violinists took heed of their advice. One pedagogue complained in 1919 that 'many players make a totally unwarranted use of the vibrato, inasmuch as they keep it up uninterruptedly'.[13] The celebrated teacher, Leopold Auer, wrote in 1921 that 'the excessive use of vibrato is a habit for which I have no tolerance, and I always fight against it when I observe it in my pupils – though often, I must admit, without success'.[14] Carl Flesch, another legendary pedagogue, recognized in 1924 that the continuous vibrato had become part of the violinistic soundscape. 'If we consider the celebrated violinists of our day', he observed, 'it must be admitted that in nearly every case they employ an uninterrupted vibrato'.[15]

It was not long before some even began to encourage the liberal use of vibrato. A 1922 German treatise included several musical examples in which vibrato is recommended on nearly every possible note.[16] By 1929 the practice had become so entrenched that one American teacher could write what would have seemed heretical nearly 30 years before: 'violin playing without vibrato', he claimed, 'is like a day without the sun – dismal and gray'.[17]

The shift from the old to the new vibrato may also be observed on record. To generalize: on recordings before 1910, vibrato – usually slight – is typically only heard decorating melodically important notes; the second decade is, in retrospect, a transitional period, with some discs demonstrating the ornamental approach and others revealing a conscious cultivation of a stronger, more frequent vibrato; and after 1920 the new vibrato is apparent in the recordings of most violinists.

A few examples should illuminate this trend. A useful starting point is with the discs left by the two most renowned violinists of the later nineteenth century – Joseph Joachim and Pablo de Sarasate. The two were stylistic opposites; Carl Flesch called them 'the two poles of the axis around which the world of the violin had turned'.[18] The Hungarian-German was considered serious, profound, intellectual, and one for whom technical matters were merely a means to artistic ennoblement; the Spaniard was admired for his elegance, suavity, his 'silken' or 'silvery' tone and perfection of technique. Yet for all their differences, they in fact shared a similar, conservative approach towards the vibrato.

Joachim militated against the overuse of vibrato, and we can hear for ourselves that he followed his own advice. In his 1903 recording of Bach's Adagio in G minor he vibrates slightly on some of the sustained tones and applies a few quick shakes to some of the highest notes in each phrase, but most of even the longer notes are played straight.[19] A more lively vibrato might be expected from his performances of Romantic music, but even on his 1903 disc of Brahms's Hungarian Dance No. 1 the old style prevails.[20] In the first 24 measures he vibrates on only nine notes and sometimes not even for the duration of a note. Vibrato does not define Joachim's sound; rather, it is a means, whether to distinguish repeated pitches, to intensify the high point of a melody, or to signal upcoming cadences.

Today, most violinists perform Sarasate's often intensely expressive works with a generous vibrato. Surprisingly, this approach is not in keeping with Sarasate's own practice. On the 1904 recording of his 'Zigeunerweisen' he plays the dramatic opening phrase with almost no vibrato, something no modern violinist would do.[21] Over the whole of the work he adds a few quick bursts to relatively short notes, applying a slower, wavering vibration to some of the longer notes. And though he tends to vibrate a bit more often than Joachim, like the latter he often plays whole phrases straight, and frequently vibrates on only a portion of a note.

A transitional phase in the use of vibrato may be discerned through a survey of recordings from the 1910s. During this period two violinists came upon the recording scene who are often cited as pioneers in the use of vibrato: Eugène Ysaÿe and Fritz Kreisler. Ysaÿe's vibrato was, according to Carl Flesch, 'a whole world away from what had been customary until then: the incidental, thin-flowing quiver "only on *espressivo* notes"'. In Ysaÿe's 1912 recording of Henri Vieuxtemps's 'Rondino' the vibrato is faster and wider than Joachim's or Sarasate's.[23] Yet while the vibrato is strong at times, at others it is entirely absent. In the beginning of the work he vibrates conspicuously only on the long notes of alternate phrases, and the difference between the vibrated and non-vibrated notes is clear. While Ysaÿe hinted at the possibility of a new vibrato, Kreisler fully realized its potential. According to Flesch, Kreisler 'started a revolutionary change by vibrating not only continuously in cantilenas like Ysaÿe, but even in technical passages'.[24] When listening to Kreisler's 1911 recording of his 'Liebesleid', for example, one gets the sense that the vibrato is not merely added to individual pitches, but that it is a vital, underlying force that connects all the notes in a legato phrase.[25] Yet while many recordings during this period point towards the new vibrato, others exemplify the old school. Recordings made around 1910 by Marie Hall or Jan Kubelik provide perfect examples of the 'thin-flowing quiver' Flesch described.[26]

The transformation from the old to the new vibrato is nicely illustrated when comparing turn-of-the-century recordings to ones made after 1920. There is a world of difference, for example, between Joachim's 1903 Bach Adagio and the 1931 and 1935 recordings of the work by Joseph Szigeti and Jascha Heifetz. In the later performances vibrato is evident on most of the sustained pitches and many of the faster ones, even some of the thirty-second notes; furthermore, both artists (especially Heifetz) vibrate faster and wider than Joachim.[27] A marked difference is also clear in later recordings of the Hungarian Dance. The 1920 and 1940 recordings by Heifetz and Toscha Seidl display a greater use and intensity of the vibrato; we hear it on 18 notes in Heifetz's recording and 25 in Seidl's – twice and nearly three times as often as Joachim. Heifetz's and Seidl's discs certainly confirm Leopold Auer's worry that he could not convince his students to resist the lure of vibrato, for the younger violinists – both pupils of Auer – vibrate almost constantly.[28]

Fig. 1: Brahms, Hungarian Dance No. 1 (arr. Joachim) Measures 1-24 (violin part only)
a. Joseph Joachim (rec. 1903) b. Jascha Heifetz (rec. 1920) c. Toscha Seidl (rec. 1940)

Similar patterns in the use of vibrato may be observed when comparing Sarasate's recordings with later ones. Zino Francescatti, for example, introduces a rich, broad vibrato from the beginning of his 1922 'Zigeunerweisen'.[29] Francescatti's vibrato is much more an organic part of his sound than Sarasate's, and is actually less variable because of its constant intensity. Interestingly, the younger violinist takes the opening much slower than Sarasate, a decision that may have been necessitated in part by his desire to vibrate on nearly every note.

Many similar examples could be chosen from the hundreds of violin recordings released early in this century.[30] A broader survey, however, would only underscore what should be clear from the written and aural evidence presented here: that the first decades of the twentieth century witnessed a fundamental transformation in the use of vibrato on the violin.

Modern scholars widely agree that early in this century violinists began to use vibrato more often and conspicuously.[31] There is, however, no consensus as to *why* this happened. A few theories have been proposed, and while some help explain the trend, none tells the whole story.

One possibility is that the new vibrato arose as a reflection of changing artistic ideals. It is important to remember, however, that the new vibrato was met with strong critical disapproval. Naturally, old-school violinists, such as Joachim and Auer, fought against the trend, but resistance continued well after the new vibrato became standard practice.[32] As late as 1950 violinist Adila Fachiri complained of the 'unremitting, nauseating vibrato used by present-day violinists'.[33] And more recently, Hans Keller decried the negative effect of the 'mania for vibrato' in modern violin playing.[34]

Some have sought to connect the trend to broader aesthetic or socioeconomic currents. In 'Vibrato on the Violin' Werner Hauck notes that 'about 1900, concepts of the Universe changed', and catalogues these changes, from impressionism and expressionism to quantum theory and psychoanalysis. 'Is it surprising', he concludes, pointing to the new vibrato, 'that violin playing, like a highly sensitive seismograph, was influenced by all this and also reacted?'[35] Hauck's conclusion, however, is not only unverifiable, but counterintuitive. Considering the sharp edges, distinct lines, and hard, polished surfaces common in the art and architecture of the time (e.g., Cubism, Art Deco, the skyscraper), one might reasonably conclude that a violin sound with *less*, not more, vibrato should arise. The strong and continued critical opposition to the new vibrato, as well as its incongruence with broader trends suggests that its development was not, or at least not solely, tied to aesthetic considerations.

Robin Stowell has suggested that the introduction of the chin rest in the early nineteenth century contributed to the rise of the new vibrato. The chin rest, which transferred the weight of the violin to the neck and shoulder, liberated the left hand, which had previously helped keep the violin firm against the chest. According to Stowell, the 'gradual adoption of the more stable chin-braced grip freed the left hand to cultivate a more fluid vibrato movement'.[36] Perhaps the full range of motion of the left hand was necessary for the new vibrato, but this explanation does not tell us why it developed in the early twentieth century and not a hundred years earlier, when violinists began using chin rests. Clearly, it was not simply the *possibility* of a prominent vibrato that led to its realization.

Another argument links the trend in vibrato to a later change in the instrument: the adoption of metal strings. After World War I, violinists began to replace their gut strings with steel ones. Metal strings have a harsher sound than gut, and violinists may have felt the need to mitigate these qualities with a generous vibrato. But the change in strings could not have been the impetus for a new use of vibrato.[37] The new vibrato arose well before most violinists were using metal strings – the switch began around 1920, but was not complete until nearly World War II. Furthermore, many violinists can be heard on record playing with a generous vibrato on *gut* strings. The change to metal, then, is best understood not as initiating the increased use of vibrato, but as contributing to the later development of the trend.

The most commonly cited force behind the rise of the new vibrato is the Austrian violinist, Fritz Kreisler. Ever since 1924, when Carl Flesch claimed that Kreisler 'started a revolutionary change' in the use of violin vibrato, critics have pointed to him as the prime mover in this aspect of performance practice.[38] Certainly the suggestion is plausible, for Kreisler was immensely popular early in this century and was widely heard not only in concert, but on his recordings, which exhibit a robust and almost constant vibrato.

There is reason, however, to question whether the origins of this trend really lay with Kreisler. There was a significant delay – at least 15 years – between Kreisler's 'revolutionary change' and the broader adoption of the new vibrato. A throbbing, nearly uninterrupted vibrato is evident on Kreisler's earliest recordings, from 1904,[39] yet while Kreisler had already been heard and admired throughout Europe and in the United States since the turn of the century,[40] most violinists did not adopt a comparable practice until the 1920s. Rather than having initiated the new practice it is most likely that Kreisler was simply held up as a model when other violinists began to use a prominent vibrato.

The question at hand thus remains largely unanswered: why did the practice of vibrato change? I will argue that to understand fully the cause of this trend we must consider a force that, not at all coincidentally, played an increasingly important role in the activities of professional violinists just prior to the change in vibrato. I propose that the conspicuous, continuous vibrato be understood as a response to the exigencies of sound recording.

A constant and strong vibrato became increasingly useful for concert violinists for whom recording came to be a way of life, and it did so in three ways. First, it helped compensate for the limitations and liabilities of early recording equipment. Second, it could obscure imperfect intonation, which is more noticeable on record than in a live setting. And third, it could offer a greater sense of the performer's presence on record, conveying to unseeing listeners what body language and facial expressions would have communicated in concert.

When recording for the megaphone-shaped acoustic horn, the violinist faced a set of unwelcome alternatives: play as close as possible to the horn and risk hitting it – thus ruining the take – or play at a comfortable distance, even at maximum volume, and still be practically inaudible to the horn. Violinist Arcadie Birkenholtz remarked on the former alternative, recalling that in the days of acoustic recording 'you had to get very close to the horn for the tone to register. And when you did that, sometimes your bow or arm hit the horn and that ended it – you had to make the record over.'[41]

A different, but equally unwelcome problem faced violinists recording with microphones, which began to be used in the mid-20s. The problem was that microphones picked up the frictional sounds of the moving bow, sounds seldom heard in the concert hall because of the distance between violinist and audience. Veteran recording artist Louis Kaufman noted that in the studio, 'You must be a little more careful with the bow pressure. You dare not press and get the extremes of *forte* that you could get in a hall in which the airspace swallows up a lot of the surface noise'.[42]

The violinist recording acoustically, then, needed a way to project sound to the horn, and not simply by playing louder. And the violinist recording electrically – with microphones, that is – needed a way to avoid projecting normally inaudible scratchiness, but without sacrificing tone or dynamic range.

Vibrato, I would suggest, helped violinists avoid the horns of both dilemmas. For those recording acoustically, a strong vibrato helped project their playing to the none-too-sensitive machines. This is due to the periodic fluctuations in intensity which accompany the use of vibrato. These fluctuations are variations in pressure that result from the contraction and expansion of air, and contribute to a sound's loudness.[43] By using more vibrato the recording artist could increase the effective loudness of a note without having to increase bow pressure unduly. This correlation is illustrated in the comparison of two acoustic recordings of Chopin's Nocturne in E-flat, arranged for violin by Pablo de Sarasate. The last note is a high E-flat, so high that it was normally unrecordable by acoustic machines. Indeed, it is almost inaudible on Mischa Elman's 1910 recording.[44] It is much more easily heard, however, on Jascha Heifetz's 1918 cut.[45] The reason has to do with vibrato. Elman uses no vibrato on the high note, and so it practically disappears. Heifetz's E-flat is audible, and is audible by virtue of the vibrato which gives it a pulsing quality.[46]

While an increased use of vibrato helped compensate for the insensitivity of acoustic horns, it also allowed violinists to project their sound to microphones while minimizing bowing noise. Louis Kaufman recognized vibrato's usefulness in recording with microphones: 'This is something of a trick, you know – getting around the surface [noise] and yet getting the intensity at the same time. The vibrato has to be somewhat heightened: it has to be somewhat faster than you really need for a public hall.'[47] Remarkably, Kaufman knowingly increased his vibrato to meet the special needs of recording. Perhaps others recording with microphones also began to find that a heightened vibrato met the demands of the new process. Indeed, many of the violinists whose recording careers spanned the acoustic and electrical eras, including Bronislaw Huberman, Mischa Elman, and Jascha Heifetz, can be heard using a stronger vibrato in their electrical recordings.

In addition to aiding in the projection of sound, an increased use of vibrato also helped recording violinists hide imperfect intonation. While mistakes in intonation are of course noticeable in a live performance, they pass quickly and may be obscured if the listener is far away or is paying close attention to the visual aspects of the performance. Poor intonation is much more easily perceived on recordings, however, because of the performer's proximity to the horn or microphone and because of the lack of the visual element. More important, recorded performances are repeatable, bad intonation and all. Violinists would certainly want to avoid blemishing their recordings with a sour note which, after repeated hearings, listeners would come to expect and dread.

So where does vibrato fit in? Violinists *not* using vibrato must be absolutely precise in terms of finger placement; otherwise, any slight inaccuracies will immediately be heard as out of tune. Paradoxically, with vibrato, one need not *be* precise in order to *sound* precise. A note played with vibrato – as long as it is not too slow or wide – is heard as a single pitch, which of course is the result of a rapidly moving left hand. Thus, as long as the violinist vibrates immediately upon placing the fingers, precise positioning is less necessary than when using a straight tone. Significantly, the use of vibrato to avoid bad intonation came to be widely recognized in the age of recording. In 1919 violinist Edmund Severn remarked that vibrato 'is used as a camouflage to 'put over' some very bad art in the shape of poor tone-quality, intonation and general sloppiness'.[48] Leopold Auer complained in 1921 that too many artists employed vibrato 'in an ostrich-like endeavor to conceal bad tone production and intonation'.[49] And in 1924 Carl Flesch noted that violinists using vibrato may 'create the impression of playing in tune'.[50]

Taken together, the heightened perceptibility of poor intonation on recordings, the broad recognition that vibrato could compensate for errant finger placement, and the concurrence of the rise of the new vibrato and the increased recording activity among violinists all suggest the following conclusion: that professional concert violinists began to use more vibrato in the age of recording in part to conceal imperfect intonation from the unforgiving phonograph.

I say 'in part' not to hedge, but to acknowledge a final way in which vibrato may be understood as a response to the exigencies of recording. I would argue that vibrato could also help compensate for the loss of the visual element in recordings. In a concert setting – unlike in a recording – performers communicate to audiences and audiences react to performers not solely through sound, but through sight as well. Consider Robert Schumann's remark about Franz Liszt in performance:

> Within a few seconds tenderness, boldness, exquisiteness, wildness succeed one another; the instrument glows and flashes under the master's hands. He must be heard and seen; for if Liszt played behind a screen a great deal of poetry would be lost.[51]

More recently a psychologist found that subjects could easily guess a performer's expressive intentions simply by watching the performance. In fact, subjects were more accurate in specifying the performers' intentions in this way than when hearing but not seeing, or hearing *and* seeing.[52] One implication of Schumann's remark and the results of the psychological study is that listeners lose a great deal of information about the expressive manner of performances heard on recordings.

I propose that the more frequent and prominent use of vibrato helped recording violinists express to unseeing listeners what their gestures and expressions could not. It is no coincidence that in the age of recording, violinists began to recognize that vibrato could help convey emotion. Early in this century vibrato was variously described as reflecting the violinist's 'innermost soul', as an 'inner, psychic vibration', or as 'the barometer of our emotions and inspirations'.[53] It is important to realize that such comments were seldom made before this time – that is, before violinists began to record in significant numbers – and would have been made about the bow, not vibrato.

Violinists saw another important function for the vibrato as well: the individualization of tone. Siegfried Eberhardt wrote in 1910 that 'the individual characteristics of different artists are recognizable only when the vibrato is employed'.[54] In 1924 Carl Flesch remarked that vibrato could even identify an unseen violinist:

> If two violinists, whose tonal qualities differ most widely, play the same sequence of tones on the same instrument behind a curtain, each using his own vibrato, the individual player may be easily and surely distinguished, while without the participation of the left hand the identity of the player can only be determined by chance.[55]

I believe that the new focus on the vibrato as both a means for conveying emotion and for distinguishing among violinists is connected to recording's missing visual dimension. When seeing always accompanied hearing in musical performance, there was no question as to the identity of the performer, and the expressiveness of a performance was strongly a function of the visual – and visceral – impact of the artist's physical presence. It was purely hypothetical for Schumann to consider Liszt (or any other performer) playing out of sight – it just was not done. But when Carl Flesch spoke of listening to two unseen violinists, he was not being fanciful, for every recording artist plays and is heard as if 'behind a curtain'.

Although there was no way for recording violinists to replace the visual dimension of live performance, it was possible to put a clearly individual stamp on one's playing and even to restore some of the 'lost poetry' through the use of vibrato. Mischa Elman's 'throbbing' could be easily distinguished from Jascha Heifetz's 'nervous' vibrato; and no one would confuse Fritz Kreisler's omnipresent shake for Marie Hall's essentially decorative use of the technique. And within a single work, any of these artists might choose to emphasize certain notes or phrases with added vibrato, or to communicate increasing or relaxing emotional intensity through changes in the speed or width of vibrations. Yet while every variety, shade, and speed of vibrato may be heard, the study of historical violin recordings clearly reveals the transformation of vibrato from an accessory to expressive violin tone to a constituent of it.

I hope to have made a compelling case linking recording and the rise of the new vibrato. First, the timing is right: we hear the beginnings of the new vibrato just at the time recording became an important professional activity for violinists. Second, through the use of vibrato violinists were able to meet challenges *exclusive* to recording: the insensitivity of acoustic horns and the problematic sensitivity of microphones; the enhanced perception of poor intonation on record; and the lack of the visual element in recording.

I should make clear, however, that by focusing on the influence of recording I am not eliminating all other possibilities from consideration. Changing tastes, the technique of particular artists, and developments in the physical aspects of the violin surely had some role in shaping the practice of vibrato. Yet these are only contributing factors and cannot fully account for the change in performance practice. I would maintain that sound recording was the most direct cause, and perhaps the only *necessary* condition for the rise of the new vibrato.

The question still remains, however, of what exactly this conclusion illuminates. Most specifically, we may now better understand the forces behind an important change in violin performance practice. From this shift arose a new, distinct violin sound, one that remains with us today.

But there are broader implications of this study as well, for it questions the received wisdom concerning the nature of sound recording. Recording has long been understood as a tool for capturing, documenting, and reifying the ephemeral art of musical performance. Yet this preservational tool is also, in fact, a catalyst. In response to the limitations and demands of this technology violinists made a small adjustment in their playing, but one with profound consequences. Moreover, not only may recording influence the way music is performed, it also has the power to shape our standards of beauty in music. As we have seen, the new vibrato arose as an accommodation to practical circumstances, but later came to be valued for its expressive potential. Necessity, it seems, may sometimes be the mother of aesthetics.

Notes

1 As early as 1677 the bow was described as 'the soul of the violin'. Bartolomeo Bismantova, quoted in D. Boyden with S. Monosoff, 'Violin Technique', in *The New Grove Violin Family* (New York, 1989), 72. Similar statements were made into the twentieth century. See L. Capet, *La Technique Supièrieure de l'Archet* (Paris, 1916), 24.

2 L. Spohr, *Violinschule*; quoted in S. Sadie (ed.), *New Grove Dictionary* (London, 1980), s.v. 'Vibrato', by R. Donington.

3 P. Baillot, *The Art of the Violin* (Evanston, IL, 1991, ed. and trans. L. Goldberg), 240-41.

4 C. de Beriòt, *Méthode de violon* (Mainz, [1858]), 242. This example is reproduced in C. Brown, 'Bowing Styles, Vibrato, and Portamento in Nineteenth-Century Violin Playing', *Journal of the Royal Musical Association*, 1988, 113, 115.

5 David and Joachim quoted in Brown, op. cit. (note 4), 114.

6 A. Saunders, *A Practical Course in Vibrato for Violinists* (London, 1900), 7.

7 Other historical surveys of violin vibrato support this conclusion. See, for example, S. Reger, 'Historical Survey of the String Instrument Vibrato' in C. Seashore (ed.), *Studies in the Psychology of Music* (Iowa City, 1932), vol. 1, *The Vibrato*, 289-304; F. Neumann, *Violin Left*

Hand Technique: A Survey of the Related Literature (Urbana, IL, 1969), 111-26; Brown, op. cit. (note 4); and R. Philip, *Early Recordings and Musical Style: Changing Tastes in Instrumental Performance, 1900-1950* (Cambridge, 1992), 97-108.

8 The Strolling Player, 'The Everlasting "Vibrato" ', *Strad*, 1908, 18, 305.

9 J. Winram, *Violin Playing and Violin Adjustment* (Edinburgh and London, 1908), 34.

10 Originally published as *Der beseelte Violinton* (Dresden, 1910). Quotation from English version, translated by M. Chaffee as *Violin Vibrato* (New York, 1911), 14.

11 P. Bytovetski, *How to Master the Violin* (Boston, 1917), 77.

12 P. Hodgson, 'Vibrato', *Strad*, 1916, 27: 148.

13 E. Gruenberg, *Violin Teaching and Violin Study* (New York, 1919), 117.

14 L. Auer, *Violin Playing as I Teach It* (New York, 1921), 62.

15 C. Flesch, *The Art of Violin Playing* (New York, 1924), 1: 40.

16 F. Rau, *Das Vibrato auf der Violine* (Leipzig, 1922), 48-53.

17 F. Hahn, *Practical Violin Study* (Philadelphia, 1929), 137.

18 C. Flesch, *Memoirs* (London, 1957), 79.

19 J. Joachim, J. S. Bach, Adagio from Sonata in G Minor, BWV 1001, Pearl compact disc, GEMM CD 9101.

20 J. Joachim, J. Brahms, Hungarian Dance No. 1 (arr. Joachim), Pearl compact disc, GEMM CD 9101.

21 P. de Sarasate, 'Zigeunerweisen', Pearl compact disc, GEMM CD 9101.

22 Flesch, op. cit. (note 18), 79.

23 E. Ysaÿe, H. Vieuxtemps, 'Rondino', Sony compact disc, MHK 62337.

24 Flesch, op. cit. (note 15), 1: 40.

25 F. Kreisler, 'Liebesleid', Biddulph compact disc, LAB 009-10.

26 Three of Hall's recordings have been reissued on Pearl compact disc, GEMM CD 9101; Many of Kubelik's discs have been compiled on *Jan Kubelik, The Acoustic Recordings (1902-1913)*, Biddulph compact disc, LAB 033-34.

27 J. Szigeti, J. S. Bach, Adagio from Sonata in G Minor, BWV 1001, Music & Arts compact disc, CD 813; J. Heifetz, J. S. Bach, Adagio from Sonata in G Minor, BWV 1001, BMG Classics compact disc, 09026-61734-2.

28 J. Heifetz, J. Brahms, Hungarian Dance No. 1 (arr. Joachim), BMG compact disc, 0942-2-RG; T. Seidl, J. Brahms, Hungarian Dance No. 1 (arr. Joachim), Appian compact disc, CDAPR 7016.

29 Z. Francescatti, P. de Sarasate, 'Zigeunerweisen', Biddulph compact disc, LAB 030.

30 For further comparisons, see Robert Philip's survey of recorded string vibrato, which complements and corroborates the present one. Philip, op. cit. (note 7), 97-108.

31 The following writers have remarked on the new use of vibrato in the first decades of the twentieth century: W. Hauck, *Vibrato on the Violin*, trans. K. Rokos (London, 1975), 17-26; O. Szende, *Unterweisung im Vibrato auf der Geige* (Vienna, 1985), 10-12; Brown, op. cit. (note 4), 110-11; Boyden with Monosoff, op. cit. (note 1), 102; J. Stüber, *Die Intonation des Geigers* (Bonn, 1989), 13-14; Philip, op. cit. (note 7), 97-108; and R. Stowell, 'Technique and Performing Practice', in *The Cambridge Companion to the Violin* (Cambridge, 1992), 130-31.

32 See, for example, H. Vercheval, *Dictionnaire du violoniste* (Paris, 1923), 129; and F. Bonavia, 'On Vibrato', *Musical Times*, 1927, 68, 1077-78.

33 A. Fachiri, 'Trends in Violin Playing', *Music and Letters*, 1950, 31, 282.

34 H. Keller, 'Violin Technique: Its Modern Development and Musical Decline', in *The Book of the Violin*, ed. D. Gill (New York, 1984), 149-50.

35 Hauck, op. cit. (note 31), 23-4.

36 Stowell, op. cit. (note 31), 130.

37 This possibility, which to my knowledge has not been proposed in print, has been suggested to me by several different scholars.

38 Flesch, op. cit., (note 15), 1: 40. See also F. Bonavia, 'Violin Playing During the Past Fifty Years', *Strad*, 1939, 50: 7; H. Roth, *Master Violinists in Performance* (Neptune City, NJ, 1982), 26; Keller, op. cit. (note 34), 149; Boyden with Monosoff, op. cit. (note 1), 102; Stowell, op. cit. (note 31), 131; and Philip, op. cit. (note 7), 106.

39 F. Kreisler, *The Kreisler Collection: The Complete Acoustic HMV Recordings*, Biddulph compact disc, LAB 009-10.

40 L. P. Lochner, *Fritz Kreisler* (New York, 1950), 56-74.

41 A. Birkenholtz, interview with J. Harvith and S. E. Harvith, in J. Harvith, S. E. Harvith (eds.), *Edison, Musicians, and the Phonograph: A Century in Retrospect* (New York, 1987), 65-7.

42 Louis Kaufman, interview with J. Harvith and S. E. Harvith, in Harvith & Harvith, op. cit. (note 41), 116.

43 See S. Reger, 'The String Instrument Vibrato', in Seashore, op. cit. (note 7), 322-3 and 330-31.

44 M. Elman, F. Chopin, Nocturne in E-flat, Op. 9, No. 2 (arr. P. de Sarasate), Biddulph compact disc, LAB 035.

45 Jascha Heifetz, F. Chopin, Nocturne in E-flat, Op. 9, No. 2 (arr. P. de Sarasate), BMG compact disc, 0942-2-RG.

46 The difference in the audibility of the two E-flats probably has little to do with improvements in recording technology between 1910 and 1918. Both recordings were made with acoustic recording horns and other than the very high notes, the sound on both recordings is comparably clear.

47 Kaufman, op. cit. (note 41), 116.

48 E. Severn, interview with F. Martens, in F. Martens (ed.), *Violin Mastery* (New York, 1919), 237.

49 Auer, op. cit. (note 14), 59.

50 Flesch, op. cit. (note 15), 1: 20.

51 J. W. Davidson, 'Visual Perception of Performance Manner in the Movements of Solo Musicians', *Psychology of Music*, 1993, 21: 103.

52 Davidson, op. cit. (note 51), 103-13. Davidson's conclusions are consistent with general studies of nonverbal communication. For a summary of some of the work done in this field, see N. Grechesky, *An Analysis of Nonverbal and Verbal Conducting Behaviors and their Relationship to Expressive Musical Performance* (Ph.D. diss., University of Wisconsin, 1985), 16-56.

53 H. Wessely, *A Practical Guide to Violin-Playing* (London, 1913), 90; M. Pilzer, interview with F. Martens, in Martens, op. cit. (note 48), 179; and Hahn, op. cit. (note 17), 137.

54 Eberhardt, op. cit. (note 10), 14.

55 Flesch, op. cit. (note 15), 1: 35.

Rebecca McSwain

THE SOCIAL RECONSTRUCTION OF A REVERSE SALIENT IN ELECTRIC GUITAR TECHNOLOGY: NOISE, THE SOLID BODY, AND JIMI HENDRIX

Introduction[1]

The guitar, like other musical instruments, has undergone a number of technological developments over its 400-year history. The goal of these technological changes has been to improve some aspect of the instrument's musical performance. The nineteenth century, in particular, was one of 'prolific invention and innovation', and a number of changes were made in guitar construction, including the introduction of such added elements as systems of horn and tubes, steel vibrators, double soundboards, and other kinds of resonating apparatus.[2] These innovations seem to relate to rather straightforward issues of volume and projection. In the nineteenth century the guitar was coming to be viewed as a solo instrument, not only as an accompaniment to song or dance, and there was concern among guitar-makers to increase volume in order to better perform in this new role in recital halls and similar venues. Increased body size as well as the variety of devices mentioned above was brought to bear on the problem of loudness.

At times as a corollary to these deliberate technological experiments there occurred changes in tonal qualities. The *tornavoz*, a conical-tube device found in some guitars made by Antonio Torres around mid-century, is said to cause 'a sound-darkening effect' and 'loss of clarity' due to loss of upper resonances. In 1866 a luthier of Valencia advertised his version of the *tornavoz* with a claim that it produced 'unheard of sonorities'. However, on the whole there appears to have been no move to experiment with these tonal oddities or to explore further their aesthetic possibilities. It seems likely, in fact, that one reason the *tornavoz* disappeared from guitar-making in the twentieth century was that the changes in sound it caused were considered undesirable.[3]

The twentieth century story of the electric guitar – latest embodiment of the technological history of the guitar – may be seen as a three-staged sequence of events. In the first stage, technological innovation applied to the familiar problem of a louder sound led to a definitive solution – the *non plus ultra* of loudness, electrification. The limits of volume would be reached once the power of electricity was added to the traditional instrument. In conjunction with the technical change of this first stage, there developed another problem, the same problem that seems to have plagued earlier volume-adding devices: the occurrence of unfamiliar, undesired, and uncontrolled sounds. Specifically, for one, with electrification came feedback, a phenomenon in which amplified notes were 'fed back' from amplifier to guitar, recycling and sustaining notes. The solution to this problem, the occurrence of unwanted sounds, was clear almost as soon as the original electrification of the instrument took place. A nonresonant, solid guitar body would reduce the possibility of feedback at high volumes. In the second stage of the technological history of the electric guitar, as the boundaries of loudness began to be pushed further and further in the realm of popular music just before and after World War II, the solid-bodied instrument was perfected and came into its own.

But it is the third stage of electric guitar history that is of most interest in the present context. In this stage, sounds formerly considered to be undesirable – tremendously sustained and distorted notes – came to be aesthetically acceptable, even important, in musical performance. Thus the solid-bodied guitar, originally conceived to eliminate feedback, came to be the instrument with which feedback – deliberate, more or less controlled – was creatively integrated into popular music. The sheer volume and the control of feedback made possible by the solid body redesigned the landscape of popular music.

In this paper, I am going to apply technology historian Thomas Hughes' concept of technological 'reverse salients' to an examination of the development of the electric guitar and its music. This concept provides an approach to the history of the electrified instrument which places it in the broader context of technological history. At the same time, the particular history of the electric guitar offers, I think, a different perspective on the reverse salient concept, suggesting one kind of effect that sociocultural and aesthetic factors can have upon technological development.

First, I will discuss the concept of the reverse salient, as presented by Hughes. Next, I will outline briefly the history of the electric guitar. Then, I will discuss the reverse salient, in the form of feedback, which occurred with electrification of the acoustic guitar. I will explain that even as critical-problem solutions were being developed which would eliminate the reverse salient, the reverse salient itself was being conceptually reconstructed in terms of musical aesthetics. The result of this socially constituted, aesthetic reconstruction would be that in some fields of music the reverse salient became a forward salient on the musical front. Finally, I will very briefly suggest some possible explanations for this reconceptualization.

I. THE TECHNOLOGICAL REVERSE SALIENT

Technological reverse salients arise, Hughes says, 'in the dynamics of the system during the uneven growth of its components.'[4] The term is borrowed from military historians, who define reverse salients as those parts of an advancing line which have fallen back. 'Having identified the reverse salients', says Hughes, 'the system tenders can then analyze them as a series of critical problems. Defining reverse salients as critical problems is the essence of the creative process. An inventor or applier of science transforms an amorphous challenge – the backwardness of a system – into a set of problems that are believed to be solvable... When engineers correct reverse salients by solving critical problems, the system usually grows if there is adequate demand for its product.'[5] Further, he notes, 'reverse salients are obvious weak points, or weak components, in a technology which are in need of further development. A reverse salient is obvious, and creative imagination is not needed to define it.' What is more difficult, and more creative, is the definition of 'critical problems': a problem, or set of problems, that, when solved, will correct the reverse salient.[6]

It also happens that the formulation of critical problems for the solution of technological reverse salients leads to the creation of new systems. The history of the direct-current electrical system offers an example: the high cost of transmission of direct current power was a reverse salient presenting critical problems that were never satisfactorily solved, but working towards solutions in the 1880s, engineers Lucien Gaulard and John Gibbs, as well as Otto Blathy, Charles Zipernowski, and Max Den (of Ganz & Company of Budapest), created the nucleus of the new alternating-current system.[7]

In the case of the guitar, loss of control of sound with electrification was clearly a weak point in this particular system. It was, as Hughes says is typical of reverse salients, obvious. In the case of the electric guitar, the critical problem – the resonance of the body on the traditional instrument – was defined by the earliest innovators. As we shall see, the solution, a solid body, was offered early, but hesitantly, and was slow to be accepted. Here, we see the power of cultural context: the guitar was – and is – an instrument carrying a heavy load of tradition, a reassuring link to the past. The tradition, the artistry, the mystique of guitar-making were all bound closely to the form of the traditional instrument, the magically acoustic box with all its resonance and beauty of wood, and these could not be lightly abandoned. (Nor have they been; the acoustic box thrives.) But no sooner had the critical problem been solved than the reverse salient itself was, in some musical realms, redefined out of existence. What had been 'noise' became 'music', and this part of the technological system of the electric guitar was no longer backward, but forward, leading the way to a new popular musical aesthetic.

I suggest that this electric guitar history shows that in addition to the 'critical problem' approach Hughes illustrates, there is another possible mechanism for dealing with reverse salients. This I call the 'social reconstruction' approach. In this model, the reverse salient is not 'fixed' in technical terms, but is eliminated by means of reconceptualization. The result of this reconceptualization is that the reverse salient is transformed into a forward salient, moving in a direction slightly askew from the original line of technological battle. New technological applications, even new technologies, may follow in its wake.

II. Brief history of the electric guitar

There exist today many popularized versions of the history of the electric guitar. These are most commonly enfolded in the broader history of the guitar in general,[8] but some recent books specialize in the electric.[9] I have recently offered a short version of my own.[10]

The facts as we currently know them are that the acoustic guitar began to be electrically amplified in the late 1920s or early 1930s. The technology for this amplification was derivative of telephone, phonograph, and especially radio technology, based on principles of magnetism and electricity which had become widely understood in the previous century. The first commercially viable electric guitar was the lap steel, an instrument held on the lap of the player which probably originated in Hawaii around the turn of the century. That the lap steel led the way in electrification was due to the popularity of Hawaiian music in the first decades of the twentieth century and the subsequent integration of the lap steel, its principal instrument, into large jazz dance bands, in the late 1920s and early 1930s. Because the performance contexts of such groups were noisy – dance halls, night clubs, etc. – and because the groups themselves were large and loud, electric amplification became a necessity. Early experiments by musicians, tool-and-die makers, and radio repairmen with mechanical resonators, microphone amplification, electrostatic pickups, and pirated phonograph parts (among others), culminated in the introduction of the 'Frying Pan' by the Rickenbacker company in 1932. In 1932 and 1934 George Beauchamp of Rickenbacker's Electro String company filed related patent applications for the Frying Pan and its electronic apparatus (Figure 1). By the time the patents were granted in 1937 the Frying Pan and its successors had spawned a host of imitators.[11]

Concurrently, virtuosos of the Spanish-style guitar began to appear, notably jazzman Eddie Lang and bluesman Lonnie Johnson. The talents and personal interaction of these two musicians of disparate background – classically educated white man and blues-grounded black man – did much to develop the possibilities of the acoustic Spanish guitar as a solo instrument in orchestra contexts.

With the electric Hawaiian guitar having established a lead position in dance bands, and such artists as Lang and Johnson demonstrating the possibilities of the Spanish guitar, electric amplification of the latter became a cultural imperative.[12] In 1936, Gibson brought an electric Spanish, the ES 150, to market. This was by no means the first electric Spanish guitar, but it had a high commercial profile and signaled the acceptance of this instrument was a *fait accompli*.[13] The ES 150, like other Spanish electrics of its day, was an acoustic box in traditional form with a pickup attached.

If the ES 150 marked the acceptance of the electric Spanish guitar by musicians and their audience, the talents of Charlie Christian and T-Bone Walker secured an increasingly prominent place for the instrument in twentieth century popular music. A Texan steeped in the traditions of blues, jazz, and vaudeville, Christian appeared in black Southwest bands during the late 1930s. There he was discovered by the ubiquitous impresario John Hammond, and in 1939 he and his ES 150 moved with flawless musical ease into the world of white middle-class national dance orchestras, playing and recording with the Benny Goodman band.[14]

Walker, another Texan and sometime partner to Christian, remained within the blues tradition and reshaped that tradition in accordance with the possibilities of the

Aug. 10, 1937. G. D. BEAUCHAMP 2,089,171

ELECTRICAL STRINGED MUSICAL INSTRUMENT

Filed June 2, 1934 3 Sheets-Sheet 1

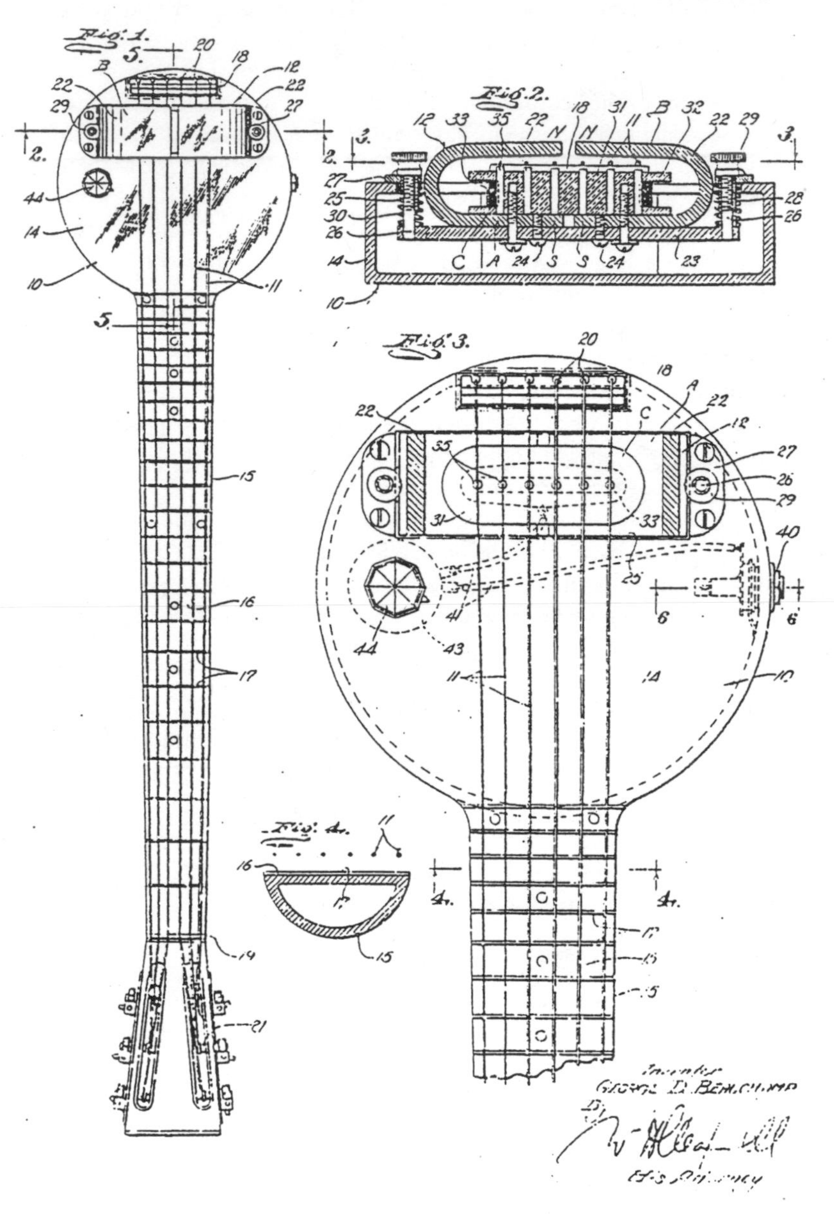

Fig. 1: Drawing from patent #2,089,171 for George Beauchamp's electric lap steel, 'Hawaiian' style guitar.

electric guitar.[15] In the hands of these men, the electric guitar began the process of its emergence into the popular music spotlight.

III. Feedback as noise, the reverse salient

Beauchamp and others recognized immediately that with electrification the acoustic qualities of traditional hollow-body guitars had become irrelevant. It was also obvious that electric amplification led to the occurrence of undesirable sounds. One of these noises was feedback, the result of amplified notes being recycled through guitar strings and pickup. This could cause a note to sustain for a long time, longer than the guitarist might wish. Feedback could also cause strange sounds, particularly when the early single-coil pickups (inadequately shielded) picked up and recycled the sounds from the amplifier as well as the sounds of string vibrations. I think there are some hints of this problem in the early patent documents.

Beauchamp's patent application of 1934 suggests that 'the body may be hollow ... to be light in weight, it being understood that in some instances it may be desirable to make the body solid'.[16] The 'some instances' Beauchamp had in mind may have related to high-volume performances. In 1936, Gary Hart of Gibson applied for a patent on 'an electrically amplified stringed instrument embodying a body member which is substantially non-resonant and means for amplifying music produced thereon to any desired degree'[17] (Figure 2). Hart's patent depicts a hollow body guitar with 'relatively thick' body walls, 'with the result that the instrument is deprived of the quality of resonance.' Thus, 'the only audible effect produced is that of the vibrating strings themselves unamplified by a sounding box effect'. The volume-and-feedback issue may have been of particular importance in the recording studio, where technology and techniques of the late 1930s and early 1940s were ill-equipped to handle the electric guitar's increased volume.[18] Charlie Christian's recorded performances from around 1940 are at relatively low volume, perhaps because of recording limitations, perhaps because of limitations imposed by feedback, perhaps due to aesthetic considerations.

It was specifically with the electric Spanish guitar that feedback problems were most troublesome. The early electric Hawaiian guitars were already solid-bodied or semi solid-bodied. This, with the possible damping effect of being played on the guitarist's lap, meant that feedback may not have been a serious problem, at least in live performance. Then there are aesthetic considerations. In a 1936 column for 'Down Beat', lap steel guitarist Jack Miller of the Orville Knapp Orchestra replied to a reader's question, 'How much volume do you use?' by saying, 'As little as possible as the tone from an electric guitar is very penetrating'.[19] Elsewhere, Miller admonishes the Hawaiian guitarist to play 'not loud but good'.[20] Another 'Down Beat' columnist recommended to lap steel players judicious use of 'power, both electrical and physical.'[21]

These factors of acoustics and aesthetics, however, did not seem to apply to the

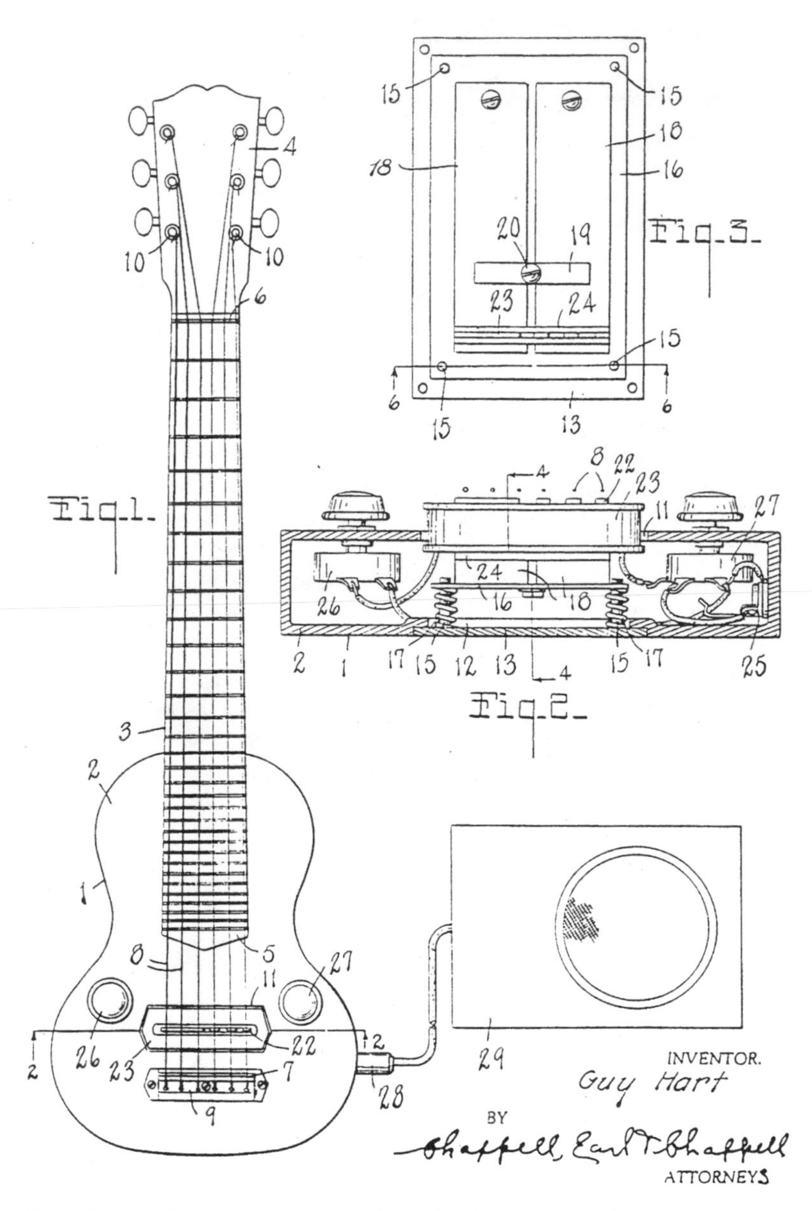

Fig. 2: Drawing from patent #2,087,106 for G. Hart's 'nonresonant' guitar.

Spanish-style guitar played by blues guitarists in noisy bars and juke joints. By the early 1940s some electric Spanish guitarists, in live performance, were certainly pushing the boundaries of loudness. One such was urban blues guitarist Memphis Minnie. She was reported as playing her electric hollow body (a National), 'amplified to machine proportions – a musical version of electric welders plus a rolling mill.'[22] For this amount of volume, a solid-body instrument would be of great value.

There were, in fact, Spanish solid bodies manufactured in the 1930s. Vivi-Tone made some backless and sideless electrics in 1933 which have been called 'essentially the first solidbody electric guitars'[23]; by about 1935 Rickenbacker had a Spanish version of its electric Hawaiian which was a semi-solid made of Bakelite;[24] and the Slingerland Company offered a solid-body electric Spanish version of their lap steel, the Songster (with an early version of humbucking pickups) in 1936.[25] None of these, however, was commercially successful.

We should note here that the original stated purpose of electrification was only amplification. The sound, it was hoped, would simply be the sound of a guitar, but louder. In his Frying Pan patent, Beauchamp says that the electromagnetic pickup provides 'true reproductions of the sounds produced by the vibrations of the strings.'[26] A patent application from Lloyd Loar of the Acousti-Lectric Company (makers of the Vivi-Tone guitar mentioned above) describes a pickup and amplifier which would transform string vibrations 'into a sound which is a true reproduction of the note set up by the vibration of the strings.'[27] Fortunately (or unfortunately, depending upon one's point of view), these claims were not entirely found true in practice.

Almost immediately there was a recognition that the electric guitar, whether Hawaiian or Spanish, was in fact a very different musical instrument from its acoustic parent. A 'Down Beat' columnist called the lap steel a 'marvelous new instrument' and a 'new voice',[28] and the editors of 'Down Beat' refer to it as 'this new orchestral instrument.'[29] The same year, 1936, in a Rickenbacker advertisement, the Electro guitars are said to bring 'New voices, new volume, a new emotion' to music, and to have a 'new tone quality, highly pleasing and entirely distinct from anything you have ever heard before.'[30] Even Beauchamp's and Loar's patents hint at some eccentric possibilities. Loar notes that his device 'permits many special adjustments to meet special occasions'; these include playing both acoustically and electrically at the same time (e.g. when pickup output is fed directly into a broadcast amplifier), and emphasizing or deemphasizing treble or bass sounds.[31] Beauchamp also says that his guitar is 'adapted to be played manually in any typical or desired manner and electrically reproduce the sound or music at a remote point.'[32] Note the potentially revolutionary phrase, 'in any ... desired manner'... Whatever Beauchamp had in mind, the latter suggestion of 'music at a remote point' conjures up for me images of bluesman Guitar Slim in the 1950s, his hair colored metallic blue or flaming red, being carried into Louisiana nightclubs on the shoulders of large men, or hanging from the rafters by his knees, all the while playing his guitar at top volume at the end of a 350-foot cord.[33]

But first, the solid-body guitar was designed according to the wishes of musicians to *eliminate* electronic feedback and maximize volume. It was with these goals that guitarists (mostly country and western players) came to Leo Fender, a radio repairman and electronic tinkerer in Fullerton, California, soon after World War II. The result of this collaboration between technician and musicians was the 'Fender Esquire', introduced in 1950. The promotional copy detailed the instrument's advances:

'Because the body is solid, there is no acoustic cavity to resonate and cause feedback, as in all other box-type Spanish guitars. This guitar can be played at extreme volume without the danger of feedback.'[34] Despite a scornful reception at the National Association of Music Manufacturers (NAMM) Chicago trade show of 1950,[35] guitarists showed the same enthusiasm for the Fender solid body that had greeted the ES 150 fourteen years earlier. The problem of feedback was largely solved, and volume levels could approach the high-end tolerance of the human auditory system.[36]

However, the process of reconceptualization was already underway. It may be that distorted or fuzzy sounds from overdriven amplifiers were the first electric guitar noises to be redefined as music.[37] But feedback was not far behind. By the early 1950s Guitar Slim (he of the 350-foot cord) was musically using feedback by means of P.A. systems and volume and tone controls on his hollow-bodied guitar.[38] He soon switched to a solid body (Les Paul, and later to a Stratocaster), which increased his control and added sustain.[39] These musicians were living and working on the periphery of middle class society, both white and black. But the colorblind, omnipresent medium of radio broadcasting, and the growing recording industry ensured that these sounds could be – and were – heard in many social milieus. And these electric guitar sounds seem to have resonated with very large numbers of people across geographic, racial, and social boundaries.

Fourteen years after the 1950 NAMM show, the Beatles released the single 'I Feel Fine' with a short feedback statement at the beginning of the song (Parlophone R5200; Capitol 5327). In 1969, Jimi Hendrix blew away the Woodstock crowd with 'Star Spangled Banner', his solid-body Stratocaster setting up howling feedback – not the rather tentative studio experiment of the Beatles, but a shocking live demonstration of the fact that the electric guitarist 'plays the guitar and the electricity.'[40] Nineteen years after the solidbody became a commercial reality, the noise it had been designed to eliminate had been definitely reconceptualized as music.

What had happened? Since the turn of the century musicians had been experimenting with heretofore nonmusical sounds and with extreme variations on standard musical tones and structures, often employing electrical technology for recording and instrumentation. These experiments had produced musics and instruments (music concrète, the trautonium, etc.) which had been of interest to some portions of the elite audience. But there does not seem to have been wide acceptance, even among the educated classes. Just after World War II, a musical instrument manufacturer complained bitterly in *The Journal of the Acoustical Society of America* that the 'musical instrument industry and musicians are extremely backward in accepting, manufacturing, and using new instruments, or improvements of old ones'. This man's company held about sixty patents on 'new electronic instruments and tone production methods.'[41] The response to this complaint was that electronic musical instruments were neither equal to nor superior to traditional instruments,[42] and were not 'designed for constructive musical purposes...'[43] In mass market terms, electronic and electrified instruments were failures. A key to this fact is that none of the instruments manufactured by Miessner and others succeeded in reconfiguring the musical aesthetics in their realms (no 'constructive musical purposes' were achieved, in the view of critics). Nor did they succeed equalling or surpassing the sounds of traditional instruments.

Yet the electric guitar was not only widely accepted after mid-century, but became a vehicle for expression of a new popular music aesthetic. With the use of controlled feedback, into the popular music mainstream came distortion and all the other myriad effects used by guitarists in several genres. The sound of popular music was ineluctably changed. Playing in 'any desired manner', as Beauchamp had put it, guitarists had realized the potential of the electric guitar as an entirely new instrument with a new sound and unexplored possibilities. The experimentation continues. (Anyone who doubts that should listen to Pat Metheny's 1994 disc 'Zero Tolerance for Silence' [Geffen DGCD 99998].) The feedback element of the unwanted noise reverse salient had been aesthetically reconstructed. But why?

IV. Who Reconceptualized the Reverse Salient, and Why?

Acceptance of electric guitar feedback (and other noises) as music seems to have begun on the periphery of mainstream American culture. That is, the penchant for ever-increasing volume, which carried musicians into an exploration of such noises, seems to have arisen in black nightclubs and white country music dance halls. While the white and black bourgeoisie argued about the relative merits of electricity in music, just at the corner of their lateral vision African-American and hillbilly musicians embraced the power that electricity gave them. These musicians were, at the beginning, free of the critical observation of Western European cultural high priests: 'no one' cared what they did. All the 'no ones' who cared, the audiences, seemed to understand and to relish what their expressive alter-egos, the musicians, were doing. Just as the great debates about jazz in mainstream musical circles during the twenties and thirties (was it 'music'? was it ruining public taste?) were irrelevant to the growing mass audience, so was the opinion of cultural high priests irrelevant in black Chicago clubs of the 1940s. Eventually, of course, that audience included the descendants of the high priests: Beatles devotees and the flower children at Woodstock.

To me it seems likely that the denizens of the periphery had hit upon a meaning or set of meanings that needed expression. That is, there was a social and individual constructive task that needed to be accomplished which could be accomplished through the manipulation of the electric guitar. After World War II, the expression of those meanings, and the accomplishment of that task, became important to the middle class cultural majority as well. The electric guitar was the perfect vehicle, the perfect tool. Historical contingency and sociocultural circumstances had made it so.

What were these meanings, these tasks? No doubt there were many expressive and constructive needs. Part of the power of the guitar, I believe, lies in its flexibility as metaphor, as sign. I suspect that on the socioeconomic periphery in this century, in America, one urgent need might have been the venting of frustration, or a temporary easing of a frustration which arose from a new awareness of limitations of access to power in the broadest sense. The ecstatic escape achieved under the influence of very loud, 'dirty', guitar music (with its boundary-breaking feedback and other sounds) may be rooted in the fleeting sense of power it bestows and expresses – in and for the group, in and for the guitar player (of one's own class and fate), in and for the individual auditor. There is no other instrument with the sheer power of the electric guitar, a power I believe is as much metaphorical as physical, as I have argued elsewhere.[44] Another meaning which might have needed expression was the necessity of reconciling modern life – its speed, urbanization, and noise – with traditional values. The acoustic guitar is an old instrument which has carried many meanings. It has always been mobile, individualistic, sexually loaded. Now, the *electric* guitar, conceptually linked with the power and modernity of electricity, has replaced the piano as the central musical expressive and symbolic vehicle for twentieth-century Americans.

Finally, I would propose that whenever we find a reverse salient being reconceptualized, as feedback noise was for the guitar, we look for a system in social flux. And we might expect the reconceptualization to originate with marginalized or peripheral people, for whom the reverse salient as it stands is of little importance – is less likely to be perceived as undesirable – because of the marginal nature of their participation in the social system with which it articulates.

As anthropologist Clifford Geertz has said of the arts in general certain activities everywhere seem specifically designed to demonstrate that 'ideas are visible, audible..., that they can be cast in forms where the senses, and through the senses the emotions, can reflectively address them. The variety of artistic expression stems from the variety of conceptions men have about the way things are, and is indeed the same variety.'[45]

Perhaps it was the variety – disparity – of conceptions about 'the way things are' between socially peripheral and mainstream Americans of the first half of the century which produced the original electric guitar, a technological system with a reverse salient of unwanted sounds. Then, later, another (or related) set of varying conceptions of reality – this time between the post-World War II generation and the preceding one – resulted in the reconceptualization of the noise of feedback as music. Thus was the reverse salient reversed, reconceptualized, making audible and addressable the ideas of twentieth-century American musicians and audiences alike.[46]

Notes

1 This is a revised version of a paper presented at the Technology and Music Session, International Committee for the History of Technology Symposium, Budapest, Hungary, August 1996. I am grateful to Dr. Hans-Joachim Braun for his gracious invitation to participate in that symposium and for his support and encouragement. Thanks to R. K. Watkins of Vintage Guitar magazine for instructive conversations and suggestions. Thanks also to Dr. Harris Berger of Texas A&M, Susan Schmidt-Horning of Case Western Reserve, Dr. Andre Millard of the University of Alabama, Dr. Charles McGovern of the National Museum of American History, and to the Anthropology Faculty of the University of Colorado for their support. I claim sole credit for all errors of fact or logic.

2 J. L. Romanillos, *Antonio De Torres, Guitar Maker – His Life and Work* (Dorset, 1987), 139-40.

3 Romanillos, op. cit. (note 2), 145.

4 T. P. Hughes, *Networks of Power: Electrification in Western Society, 1880-1930* (Baltimore, London, 1983), 14.

5 Hughes, op. cit. (note 4), 14-15.

6 Hughes, op. cit. (note 4), 22.

7 Hughes, op. cit. (note 4), 86-91.

8 K. Achard, *The History and Development of the American Guitar* (Westport CT, 1990); T. Bacon, P. Day, *The Ultimate Guitar Book* (New York, 1992); T. Evans, M. A. Evans, *Guitars: Music, History, Construction and Players From the Renaissance to Rock* (New York, 1977); T. Wheeler, *American Guitars: An Illustrated History* (New York, 1990).

9 G. Gruhn, W. Carter, *Electric Guitars and Basses: A Photographic History* (San Francisco, 1994); P. Trynko, *The Electric Guitar, An Illustrated History* (San Francisco, 1995).

10 R. McSwain, 'The Power of the Electric Guitar', *Popular Music and Society*, 1995, 19: 21-40.

11 Gruhn, Carter, op. cit. (note 9), 10; Wheeler, op. cit. (note 8), 332.

12 M. B. Schiffer, 'Cultural Imperatives and Product Development', *Technology and Culture*, 1993, 34: 98-113.

13 Wheeler, op. cit. (note 8), 131-2.

14 S. Britt, *The Jazz Guitarists* (Dorset, 1984), 11. J. Sallis, *The Guitar Players: One Instrument and Its Masters in American Music* (Lincoln, London, 1994), 97-120. G. Schuller, *The Swing Era: The Development of Jazz, 1930-1945* (New York, 1989), 563-78.

15 Sallis, op. cit. (note 14), 155-72.

16 *Patent #2,089,171*, 1937, 1.

17 *Patent #2,087,106*, 1937, 1.

18 C. Ginell, *Milton Brown and the Founding of Western Swing* (Urbana, Chicago, 1994), 173. R. Palmer, 'The Church of the Sonic Guitar', *The South Atlantic Quarterly*, 1991, 90: 656-57.

19 J. Miller, 'Steel Guitar has Place in Modern Orchestra', *Down Beat*, June 1936, 13.

20 J. Miller, 'The Art of "Fill-Ins" on Electric Guitars', *Down Beat*, November 1936, 15.

21 B. Mulcahy, 'The True Value of the Electric Guitar', *Down Beat*, October 1936, 14, 18.

22 Langston Hughes, quoted in P. Garon, B. Garon, *Woman With Guitar: Memphis Minnie's Blues* (New York, 1992), 54.

23 Gruhn, Carter, op. cit. (note 9), 45.

24 Wheeler, op. cit. (note 8), 335.

25 Gruhn, Carter, op. cit. (note 9), 49.

26 *Patent #2,089,171*, 1937, 1.

27 *Patent #2,025,875*, 1935, 2.

28 Mulcahy, op. cit. (note 20), 18.

29 Edit. Note, *Down Beat*, April 1936, 5.

30 'Important Contributions to the Field of Amplification', *Down Beat*, April 1936, 5.

31 *Patent #2,025,875*, 1935, 2.

32 *Patent #2,089,171*, 1937, 1.

33 Palmer, op. cit. (note 18), 663-4; D. Wilcox, B. Guy, *Damn Right I've Got the Blues: Buddy Guy and the Blues Roots of Rock and Roll* (San Francisco, 1993), 20-1, 33.

34 A. R. Duchossoir, *The Fender Telecaster: The Detailed Story of America's Senior Solid Body Guitar* (Milwaukee WI, 1991), 10, 11.

35 Duchossoir, op. cit. (note 34), 12; Bacon, Day, *The Fender Book* (San Francisco, 1992), 13.

36 An interesting point of comparison between the synthesizer (Pinch and Trocco, this volume) and the electric guitar is that the successful instrument was produced by a nonmusician who was able to listen carefully to and understand the wishes of professional musicians. In the case of the guitar, Fender was much the same kind of man as Moog: a nonmusician technical expert interested in mass production. Moog's comment about 'responding to demand' is something Fender would certainly have said about his own shop. One is then tempted to extend the parallels to other figures in early electric guitar history, for example, musicians like Beauchamp and Paul Bigsby (an early experimenter with solid-body design), who might be considered analogous to Buchla in the synthesizer story.

37 Palmer, op. cit. (note 18), 658-61.

38 Palmer, op. cit. (note 18), 664.

39 Palmer, op. cit. (note 18), 665; Wilcox, Guy, op. cit. (note 33), 22.

40 M. Lydon, E. Mandel, *Boogie Lightning: How Music Became Electric* (New York, 1974), 155.

41 B. F. Miessner, 'Electronic Musical Instruments', *Journal of the Acoustical Society of America*, 1947, 19, 996.

42 A. R. Rienstra, 'Electronic Musical Instruments', *Journal of the Acoustical Society of America*, 1948, 20, 550.

43 H. L. Robin, 'Electronic Musical Instruments', *Journal of the Acoustical Society of America*, 1948, 20, 345.

44 McSwain, op. cit. (note 10).

45 C. Geertz, *Local Knowledge: Further Essays in Interpretive Anthropology* (Basic Books, 1983), 119-20.

46 As discussed by Pinch and Trocco, the synthesizer seems to present a kind of opposite example to the electric guitar in terms of 'interpretive flexibility': In the case of the synthesizer, 'closure' meant a narrowing-down of both structure and function of the machine/instrument, while in the case of the electric guitar 'closure' was a vast expansion of the original purposes of electrification. I suspect the key to this difference lies in the alliance of the synthesizer with the keyboard, imparting a conservative tendency to its development, as I think can be inferred from Pinch and Trocco's own description of events.

Helga de la Motte-Haber
Soundsampling: An Aesthetic Challenge

1. The Extension of Material, a Basic Concept of Occidental Music

The rattling of chains, clanging of a blacksmith's hammer, cracking of whips, ringing of cow bells, cuckoo calls, barking, howling, grunting, hissing, screeching and neighing:
these human and animal sounds – what Michael Praetorius called 'the awful and horrible music of the infernal hot chapel'[1] – long constituted something very different from traditional music. Yet these sounds have had a place in art. In the Middle Ages no mystery play could do without the depiction of the devil and centuries later devils were still part of the opera stage. But generally, these devils were not allowed to sing. They whistled instead, because whistling was not regarded as music, but as an activity of everyday life. These unpleasant everyday noises, this whistling and rattling, however, must have had some attractions; otherwise composers would not, under whatever pretext, have tried to integrate them into their works.

Already in early musical history Western composers were keen on extending musical material. But for centuries it was taken for granted that only sounds with definite pitches constituted music. The sounds of the Turkish Janissary bands imitated by Mozart and Beethoven evoked the exotic but were not heard as noise. Or should one, considering the music of Vivaldi or Beethoven, regard the calls of cuckoos or nightingales as extensions of musical material? Probably the explosion of musical material at the turn of the 20th century cannot be explained without reference to earlier attempts to incorporate noise into the world of pure tone.

Humming, thundering, bursting, clattering, tumbling, roaring, whistling, hissing, puffing, . . . moaning, bellowing, howling, laughing, groaning, sobbing. These noises come from a source different from that mentioned at the beginning. They are noises which Luigi Russolo called for in his futurist orchestra. As he pointed out in his manifesto of futurist music in 1913, he regarded the tone colours of the existing musical tones as too limited. 'Even the most complex orchestras basically have only four or five types of instrument, varying in timbre: string instruments, winds in metal or wood, and percussion. So, modern music has reached a dead end and tries in vain to create new possibilities of sound. We have to go beyond this narrow circle of pure tones and include the infinite variety of sounds and noise'.[2]

The extension of music by noise which seemed to be important in sonatas for aeroplanes or for the musical depiction of the awakening of the city had, however, to be created by newly built machines (intonarumori). They imitated what would nowadays be retrieved from a digital sampler. As often happens in musical history, the artists' intentions preceded the possibilities of technical realisation.

Futurist noise music aimed at a musical revolution. According to Russolo, 'for many years Beethoven and Wagner shook our nerves and hearts. Now we are satiated and we find far more enjoyment in mentally combining the noises of trams, backfiring motors, carriages and bawling crowds, than in re-hearing, for example, the "Eroica" or the "Pastoral" '.[3]

Noise defined a changed relationship between art and reality. Art was now to include everyday life. But the futurists' attempts to do so seem rather amateurish. Nevertheless, the futurists influenced composers outside the movement, such as Stravinsky and Varèse, who wrote shocking, intensely dissonant works. But in this process, the futurist ideal of advancing everyday noise to the status of music was narrowed. Stravinsky and Varèse merely aimed at including new sound material into music. 'Why', Varèse complained, 'do the futurists have to catch all those trembling utterances of every-day life?'[4] The idea of an extension of material – noise instead of tones – was perfectly compatible with the traditional concept of art.

Technology did continue to be a theme in music even after the futurists. But it was then regarded mainly as a means to achieve new sound realisations and was not an issue demanding aesthetic reflection. Everything which served the same purpose was of equal value. Therefore the early forms of soundsampling and the new musical instruments, which were invented shortly afterwards, are of similar importance. In 1942, John Cage talked about the howling of dogs stored on a gramophone record, which, by variation of the turntable's speed, served to obtain a gliding tone. It is possible that the anecdotal accounts about Varèse experimenting with the speed of turntables refer to the same studio. But the Ondes Martenot and the Theremin also produce gliding tones. Composers obtained new sounds by using sampled materials as well as by producing original sounds.

In the beginning of musique concrète, specific characteristics of soundsampling were of no importance. Such a specific characteristic is the possibility of combining the separation from the 'here and now' with the separation of a sound from its source. Basically, sound sampling implies a new relationship between art and fragmented reality. At first, those issues did not turn up and all technically produced acoustic phenomena were found fascinating. But this was generally true only until the 1950s. Electrically operated sound producing instruments can be regarded as precursors of the electronic music of the fifties. The Cologne electronic studio was equipped with a trautonium, which was developed in the 1920s. In its early days, this studio was concerned not only with the creation of new sounds and later, the importance of new sound generation decreased even further. After Herbert Eimert had defined electronic music,[5] the main problem in this studio was to find rules for the development of a new structural order which extended to the frequency spectrum. Inadvertantly, the young composers of the 1950s and the inventors of new instruments in the 1920s shared the premise that an extension of musical material was only important if it would lead to structurally new music. Already in 1907 Ferruccio

Busoni, prompted by the invention of the dynamophone, had hoped that composing with third tones would become possible and could lead to a reformation of music.[6] Many composers intended to replace the tempered system. Although Jörg Mager's spherophone was mainly used for theatrical sound effects, its inventor had built it for quartertone compositions.

In the twentieth century two different conceptions of the extension of mechanically produced and sampled sound material were developed. In this context material means the sum of all acoustic phenomena, but it also denotes a new system of ordering sound. Such a system had to have the same cogency as the tonality which it tried to replace. The fact that in the 1950s a conceptual conflict developed, was to a large extent due to Adorno's influential definition of material as a new compositional order and not as new sound. I will come back to this later.

2. *Material sampling and montage*

Recording of acoustic material goes back to the nineteenth century. In 1877 the phonograph was invented, followed in 1888 by the gramophone. After this, the 'now' of a sound or a piece of music could be transformed into a 'here'. Surely, in the late nineteenth century people did not realise the extent of this dramatic alteration of reality. They just enjoyed the feeling of being in a ballroom any time they liked or listening to Caruso at a price far lower than at the opera house. The reproduction of reality by technological means was only analysed after radio and sound film had caused a change which could no longer be ignored. Walter Benjamin's essay of 1936 'The Work of Art in the Age of Technical Reproduction' became particularly influential. According to Benjamin the loss of the 'here and now' is compensated by the utopia of a ubiquitous availability of the work of art. He had in mind new, more diffuse forms of perception which correspond to the montage-like views of the world.

For a long time sampling, reproducibility and montage were generally seen as being closely related to each other. The technology to handle sampled objects or sounds was a direct outcome of dealing with film. But it was also stimulated by the method of collage, developed in painting and poetry. An early example of sampling and the montage of sound phenomena inspired by film is Walther Ruttmann's sound piece 'Weekend' (1930). Prompted by the newly developed sound film, Ruttmann created a counterpart to his famous film 'Berlin – Symphony of a City'. In 'Weekend', speech, noises and music are integrated as equal montage elements. In it he created a study of the world around him which is more complex than the pieces which could have been produced by contemporary radio. Ruttmann did not find it difficult to create a musical piece from sampled acoustic events, because earlier, by montage, he had already put the images of Berlin together to form a symphony. Although there are nowadays a large number of relevant examples, Ruttmann's name should still be mentioned. His was a pioneer achievement which influenced, among other works, Pierre Henry's 'La Ville', a musical treatment of the city.

3. *MATERIAL MONTAGE AS A CAUSE OF AN AESTHETIC CONFLICT*

Early attempts to simulate sound worlds by sampled events of the human environment have first and foremost to be associated with the name of Pierre Schaeffer. His noise études ('études de bruit') are the first examples of the so-called 'musique concrète', developed around 1948/49. As with Ruttmann, Schaeffer's starting point was concrete sounds, which were recorded on the endless grooves of phonographic records. In 1948 he composed his railway étude. Rhythmic montages of railway noise were transformed into music, which no longer could be notated in a traditional way. As an example of a novel sound world, his 'Sinfonie pour un homme seul' has become almost legendary. At the Donaueschingen Festival of 1953 a clash with those composers arose who employed electronic means exclusively. The only reason why the Paris composers' piece 'Orphée 53' was not drowned in protest was that Pierre Henry, who collaborated with Schaeffer, turned up all the controls.

In view of the fact that electronic music and musique concrète amalgamated soon after this, the question is why this conflict arose at all. It probably had to do with different concepts of sound material. Composers in electronic studios, especially in Cologne, had hoped that sound synthesis could generate a new order which extended the serial principle to the structure of partial tones. Musique concrète, on the other hand, applied the concept of material to be found in surrealism. By creating montage from the accidentally found and stored *objet trouvé*, composers aimed at a novel interpretation of a world, in which the objects found in a normal human environment obeyed a different, better logic. The conflict was rooted in different aesthetic positions and focused on the use of technical apparatus. It dissolved pretty quickly. In his 'Gesang der Jünglinge' (1956) Stockhausen used a boy's voice together with environmental material recorded on tape, as well as synthetically produced sounds. What was sampled no longer had the surrealist odour of *objet trouvé*. Soon after the invention of magnetic tape it had become clear that sound synthesis would not be developed so rapidly as to soon reach the complexity of recorded material. In the visual sphere, a comparable development has recently taken place with the optical scanner. In many cases the sampling of images has proved more convenient than producing them synthetically.

In musique concrète first no manipulations of the sampled sound objects existed. Placed into a new context, they obtained a new meaning. But the technical and musical development – around 1960 in the direction of live electronics – emphasised a variable shaping of sound. After the Second World War, the ideals of surrealism had dissolved rapidly. Nobody believed any longer in the possibility of a creation of the world by montage, which the early adherents of sound sampling had hoped to attain. But changes were due. Referring to his 'Telemusik' of 1966 Stockhausen talked about 'modulations' and explicitly dissociated himself from the montage and collage concepts of the first half of this century. In 1969 he said that 'Telemusik' had become the start of a new development, by which the use of collage during the first part of the century had slowly been overcome. Telemusic is no longer collage. Instead, by modulation between 'found' objects and new sounds created by modern means, a higher unity was reached: a universality of past, present and future in countries and areas far distant from each other.[7] Stockhausen described the kind of modulation used in his 'Telemusik' in the following way: 'I modulate the rhythm of one event with the volume curve of another. Or I modulate electronic chords, which I produce myself, with the volume curve of a priest's chant, and then these with the monotonous song of a Shipibo song, etc. By these means I have, in 'Telemusik', for the first time combined music by others with my own. Until then, I had in my compositions not accepted music which already existed. But I wanted to get out of this system of exclusiveness. And in the same way, in which I have always integrated sounds which occur in life (not only from artificially built instruments), I also try to put all kinds of music, be it 3000 years or just one day old, into a new relationship with each other. We know that time differences are artificial and that everything exists in our consciousness simultaneously'.[8]

It is important to stress the visionary element in this piece of 'world music' which is expressed by the terms integration and universality. The idea of world music is anti-nationalistic. From the dissolution of the 'here and now' following reproduction it attains the utopia of a higher unity.

4. *The dissolution of sound from the 'here and now'*

Loud neighing, indignant growling, noises of different origin, shrieking, shooting, blows against iron bars: this list arises from a context which again differs from the one mentioned at the beginning. They come from the 154 kinds of sounds which Mauricio Kagel produced for his film-sound play 'Soundtrack' (1975) and they constitute one of the four structural layers of the piece. Horse neighing is accompanied by piano playing, an incessant, boring Alberti bass. 'Soundtrack' combines the acoustic happenings in the life of a family who watches a Western, with the sound of a child practising piano in the next room. Each member of the family speaks: the father: 'I wouldn't be surprised if he shoots him down soon! There they come, get at him! Idiot.' Mother to son: 'What's up? You think you are a real know-all. Practise first.' Grandmother: 'Ah, here is my tonic.' Mother: 'We are only just at the beginning and two have been shot down already.' Daughter: 'Thank you.'

'Soundtrack' shows how in an electronically mediated world family ties dissolve. This happens unnoticed, because Kagel conceives the monologues as dialogues. Also, by the installation of a sound carpet from outside in a living room, this piece hints at elements of a fragmented acoustic reality.

5. *The necessity to reflect one's subjective position*

Storage devices like film, video, records, tape or computer, distort man's natural environment. 'The sunset delivered free of charge', was Paul Valéry's ironic comment on the omnipresence of everything technically reproducible. It seems, however, that at the turn of this century a shift in the relevant aesthetic issues took place. I sketched earlier a concept of music which aimed at excluding common reality. But composers adopted devilish rattles, cracks and the squeaks and whistles of nature. It was partly the attempt to include as many elements as possible into their compositions which made them use unusual sounds, but it was also the fascination to venture into an imaginary soundscape. It was not the technological possibilities which generated this way of thinking. This is borne out by the noise machines which were built only for musical purposes and were only less-than-ideal solutions. But technological possibilities have accelerated this process.

Nowadays art makes use of different technological means. All kinds of electro-acoustic music bear this out. On a wider scale, however, there is the challenge to reflect the aesthetic changes of reality brought about by technological development. It is, indeed, necessary to mention futurism, an art movement, which, because of its later associations with fascism, became somewhat suspicious. For futurism, technology was not only regarded as a means to an end, but as something, which had changed humans' conditions of life and sense of time. In the 1960s, art was also often a reflection of the human environment; artists pointed out solutions and utopias. Stockhausen's 'Telemusik' is an example. Others, like Luigi Nono's 'Fabbrica illuminata', could also be mentioned. This piece, too, is based on sampled material, in this case on factory noises. Transformations and artistic reflections generally imply the hope of change.

Today the problem that technology has altered our natural environment, that it has broken it up and has made it accessible to us only in certain aspects, is so conspicuous that technology can no longer be regarded as a means of change. The traditional conditions of music production and reception are no longer valid. Everything electronically stored can be installed everywhere. Mozart in the living room, who would find this disturbing except those who do not like classical music? It was John Cage who reflected on this situation very early. The score of his 'Credo in US' of 1942 for percussion quartet, radio and a phonograph has details on the use of records and the radio. The phonograph player is instructed to play something classical, for example Dvorak, Beethoven or Shostakovich. Every classical composer, stored on record, can now be transposed into each concert. Cage's piece does not offer a solution. At the end, the record drowns the live players. But in this early media composition he demonstrated a new creative use of records and radio.

In view of the abundance of material, one of the main problems an artist faces today is the decision about the choices he or she should make. This is, indeed, an urgent problem. But it is only part of the question about the changes which technology has brought about in our perception. Perhaps the objective of art should be to find answers to the question of how to make perception more sensitive.

Sampling offers a good example of illustrating those changes in perception. Normally, signs are signals of an object, of a sound source, they are its properties. A large, hollow body sounds low and muffled. If sampled, it becomes a sound object.

Pierre Schaeffer called this an 'object sonore'.[9] But those sampled signs partly retain the properties of the object from which they originate. These are mainly affective properties: sounds can be charming, threatening, alarming or enticing. Especially in the 1960s these associations made sound compositions attractive, because they made it possible to regain expression. However, sounds as sampled objects are no longer signs. The original reference has dissolved. They can be linked with many other visual or acoustic objects and then attain a new and perplexing symbolic quality. Detached from their 'here and now' they can, now in a sampled manner, be placed into different contexts. They can also serve as acoustic ready-mades. When one thinks of the innumerable interpretations of Duchamps' objects, detached from their original context, one will soon get an idea of the importance of context, which is, of course, partly determined by the objects themselves. The meaning of a sign results from a network of relationships, the sign has no meaning in itself. In sound sampling the 'here' of a sign is reduced to its time character of the 'now'. Its place of origin can no longer be detected, although there are vague associations to it. In Bill Fontana's installations sounds are transmitted round the earth via satellite. Maybe frogs from the bank of the Danube will croak in the centre of Vienna. This happens by direct transmission in real time and, for Fontana, is very important, because he can only very seldom be persuaded to use sampled sounds. I mention this example, which conceptually transgresses sampling, because in it the places of origin are not only dissolved, but in the dissolution of the 'here and now' the time coordinates are also lost: the 'now' of the sounds is at a place different from the 'here'.

Technology creates a world full of simulated events which dissolve the space-time coordinates of the subject. We should not forget, however, that this also applies to the everyday world. Being transferred to a different place and time by watching television means that one is already part of a network of artificially created meanings. The coordinates, which determine the subjective present, are thereby dissolved. Spatial distances shrink to zero. There is a compression of time, and the technologically determined fictive worlds make it difficult to establish a 'here and now' for ourselves. The more the events around us can be simulated, the more it is incumbent upon the arts to enable the perceiving subject to reflect on its own reality.

The relationship between art and life has altered. The function of art to uncover basic principles of life will decrease. Instead, art will have to create irritations and make a process of perception possible which is necessary for determining one's own subjective position. Bill Fontana's installations show that artists feel called upon to create such irritations. In his installations dogs bark from church spires and rivers run down the walls. Patterns of perception are broken up and this should induce the recipient to reflect on his own position. Today, art has become an experiment of consciousness. It is probably easier to pursue this in one of the new technically determined genres than in traditional concert music. But perhaps even there something may be detected which, with comparable cogency, brings up the issue of the position of the individual in a world which is dissolved in a multitude of variable sign constellations. In an experiment of consciousness, art irritates and the recipient is obliged to find his or her own frame of reference. But this is different from knowing an absolute frame of reference. Perhaps we should no longer talk about the issue of right or wrong, but try not to get lost in the complex and variable network of signs called reality.

Notes

1 M. Praetorius, *Polyhymnia caduceatrix* (1619, complete edition of his works, Wolfenbüttel, Berlin, 1935), Vol. 17, VIII.

2 L. Russolo, 'L'arte dei rumori' (1913), quoted after L. Schulenberg, W. Bortlik (eds.), *Drahtlose Phantasie* (Hamburg, Zürich, 1985), 23.

3 Russolo, op. cit. (note 2), 24.

4 See L. Hirbour (ed.), *Edgar Varèse: Ecrits* (Paris, 1983), 24.

5 H. Eimert, 'Vorwort zur elektronischen Musik', *Die Reihe*, 1955, 1: 7.

6 F. Busoni, *Entwurf einer neuen Ästhetik der Tonkunst* (1907, ed. Leipzig, 1916).

7 K. Stockhausen, *Texte zur Musik* (Cologne, 1969), Vol. 3, 76.

8 Stockhausen, op. cit. (note 7), 77.

9 P. Schaeffer, *Traité des objects musicaux* (Paris, 1966).

MARTHA BRECH

NEW TECHNOLOGY – NEW ARTISTIC GENRES: CHANGES IN THE CONCEPT AND AESTHETICS OF MUSIC

Throughout the ages technical developments and inventions have had a great influence on life in general, but also on music and musical instruments. However, no technological innovation has led to such radical change in music as the application of electricity. It was only about one hundred years ago that the first patents on the production of sound by means of electricity were awarded. Today both 'art' music and pop music are completely different from the music composed or played before the turn of the century. Not only has the sound of music changed, but also its conceptualization. Music is now composed which can only be played or heard by means of electricity, and new artistic genres have emerged which are based upon both the new technology and a change in musical aesthetics. This article tries to present and analyse the most important steps in the development of this kind of music and the origins of new artistic genres.

It is difficult to determine whether the new technological inventions led to new art forms, or the artists and philosophers who first conceived this music gladly applied new technology to turn their musical ideas into reality. But the development was probably parallel. When Thaddeus Cahill built the first electrical musical instrument, he had no idea of the new kind of music to be played on his 'Dynamophone', but was only thinking of new possibilities in performing the music of the day. The Theremin, invented in 1920 by Lev Termen, was only known to the public as an electrical musical instrument which played well-known music in a rather uncommon way, and the same can be said of many other electrical musical instruments of the twenties. However, composers soon began to write music especially for those instruments. In order to promote the Theremin, Schillinger composed his 'Airphonic Suite for RCA Theremin and Orchestra' in 1929. Olivier Messiaen, André Jolivet, Darius Milhaud and others composed original music for the Ondes Martenot, invented 1928, and Paul Hindemith not only wrote music for the Trautonium which was first presented to the public in 1930, but also encouraged Friedrich Trautwein to extend certain features of this instrument[1] in order to facilitate intonation. Eventually, Theremin, Trautonium and Ondes Martenot were applied in mixed ensembles together with acoustic instruments. In new compositions, the instruments extended well-known timbres. However, composers never confronted electrical and traditional sounds in such a way that, for instance, extreme noises could be heard. They did study the special playing techniques and new timbres of these instruments and wrote parts which could not be played on traditional ones, but the aesthetic background of these modern looking compositions was still very traditional.

Nevertheless, there was the beginning of a new aesthetic. At the time when the first electrical musical instruments were built, many composers and artists dreamt of a new kind of music, one which reflected modern life and consisted of noises, new timbres, a completely different scale, or even of light and colours connected with sounds. Most of these ideas remained utopian. Ferruccio Busoni, for instance, had read of Thaddeus Cahill's Dynamophone and gave a description of it in his 'Sketch of a New Aesthetic of Music'. He looked forward to future generations controlling this unusual material,[2] the new sounds offered by this instrument. Arnold Schönberg was equally vague, writing of future music which might consist of a timbre-melody in the last short chapter of his 'Harmonielehre' (1910). A few years later, in 1913, the Italian Futurist Luigi Russolo built his Intonarumori, an orchestra of noise instruments, according to his own aesthetics of Futurist music, and in 1917 Eric Satie invented his 'musique d'ameublement', music that was to furnish a room and was not supposed to catch the attention of the people in it.[3] But these first manifestations of a radical new aesthetics of musical art were based on traditional mechanical settings and did not yet use electricity.

Today, Walther Ruttmann's 'Weekend', produced in 1930, is generally regarded as the first piece of audio art that combined a new aesthetic with new electronic technology. Ruttmann, a film director, recorded his artistic representation of a weekend on film (Lichtton), the first media which allowed the cutting of sound with precision. 'Weekend' is a documentary collage of sounds. One might call it an early example of 'musique concrète', but the strong narrative element in it does not justify the term music. It is a mixture of music, because it is made out of sounding material (identifiable noises, spoken words and some tunes) and drama or literature, because those realistic sounds tell a story. The reactions of the press show that it was recognized as an outstanding production. But 'Weekend' was not the only acoustic documentation produced at the time. In order to develop a special art form for the radio, 'Hörfolgen' (listening series) were established in German radio stations in 1928. Unlike 'Weekend', they were produced and broadcast live, but owing to the lack of recording equipment did not survive. At least some press reviews provide us with some impressions of this new art, made especially for the new 'electrical instrument', the radio. City life and reflections of modern times were the most fashionable themes of these acoustic collages. They combined poems, songs, dancing tunes and all kinds of music; literary texts and prerecorded noises, creating a rather dense acoustic atmosphere aimed at the entertainment of listeners. Music played an important role in these revues, equal to the role of texts and dialogues, and was not purely functional like the music in radio plays. Not all of the 'Hörfolgen' were revues. Text and music (or sound) had an equal role in 'Hörfolgen', and productions varied considerably. 'Radio opera' would have been a better term for some of them. Different terms existed like 'radioplay', 'Hörfolge' or 'lyrical suite',[4] given to 'Leben in diesen Zeiten' ('To Live in these Times'), written in 1929 by Erich Kästner, an important German poet and novelist, and Edmund Nick, head of the music department at the Breslau radio station. They make clear that there was a mixture of two genres in this production and there were problems, at least for the journalists, in categorizing this new art form. 'Leben in diesen Zeiten' became one of the most celebrated productions. During the following years, it was often repeated, later rewritten and became a successful stage production. Less widely successful, but equally irritating to the critics, was 'Glocken' ('Bells') by Genö Ohlischlaeger and Walter Gronostay which was first broadcast a few weeks earlier than 'Weekend'. All kinds of bells were presented acoustically in this 'Hörfolge' together with reports or small scenes spoken by actors. As one critic wrote, sound no longer served only in the background, but 'became a dramatic event itself. Here a new radioplay was created which is built homogeneously on acoustical facts and is yet based on conditions which are, in their abstractness, completely different from a musical composition.'[5]

Although the new acoustic radio genre 'Hörfolge' enjoyed popular reception and many of its works were favourably reviewed, this genre was never really considered 'art'. In a first enthusiastic review of 'Weekend' a critic explicitly points out that the 'listening montage may not be judged as a work of art'.[6] Obviously, the traditional aesthetics of 'high culture', in which the production of art for art's sake had the highest rank, was still overwhelming to this critic, who did not even use the term 'Gebrauchskunst',[7] a term generally given to any kind of radio art, because its main aim was to 'merely' entertain radio listeners. So one might conclude that this new art form irritated all aesthetic standards. However, 'Weekend' and other examples of 'Hörfolgen' seem to represent the first step towards the creation of a new artistic genre. Combining elements of formerly separated genres it made use of the opportunities of electronic technology.

Experiments with art produced especially for the new medium of radio were not restricted to Germany. But it seems as if the people in charge of radio stations in other countries showed little interest in supporting these experiments. An article in 'Der deutsche Rundfunk'[8] reported that this was clearly the case in France, where, due to lack of interest by radio officials, radio art experiments had to be made privately in the Paris home of Paul Jermée, head of a group of radio artists. According to the report, the setting of these experiments must have been the same as in public radio art 'performances' in Germany: a microphone was installed in one room and a loudspeaker in an adjacent one.[9] Additionally, the author describes this kind of radio art as comparable to the productions of F. W. Bischoff, director of the Breslau radio station and famous author of innovative radio plays and of 'Hörfolgen'.

Some years later, the next important step towards a new artistic genre also occured in Paris, although no direct link can be established between this and the experiments of the early thirties. Around 1943 a group of people developed a new concept of radio production, and soon after the liberation in 1944 began broadcasting new radio plays in the so called 'Studio d'Essay' of the French Radio. A member of that group was Pierre Schaeffer, who considered sounds, noises, spoken literature and music as equally important. In 1948, when he composed his 'Concert de Bruits', his experiments led to the birth of 'musique concrète'.[10] So again, it was the medium of radio that allowed the development of a new art, and another radio station, the

'Nordwestdeutscher Rundfunk' in Cologne, was also the home of the 'elektronische Musik' Studio which was founded there in 1952.

The two styles had very little in common, although both used amplifiers and loudspeakers. Composition rules were so different that they might be called antagonistic. Herbert Eimert, director of the Cologne studio, even felt a kind of rivalry between the two schools of electro-acoustic music in their formative years. 'Musique concrète' was made from all kinds of prerecorded sounds and noises, including those of daily life. They were electronically manipulated and put together in the studio with different compositional ideas which often took the sounds themselves into consideration. As in acoustic music, the basic sounds used in 'musique concrète' were, from a physical point of view, already complex, while the sound sources of 'elektronische Musik' were simply sound generators or basic modulators. Complex sounds and the composition itself originated from single sound cells with serial construction rules, because Eimert was a dedicated follower of this system. Stockhausen's famous 'Studie II', composed in 1954, is a good example of 'elektronische Musik', combining the serial avant-garde structure with a constructive principle (25th root of 5) that is completely uncommon in the mathematics of music. In this piece Stockhausen showed in an abstract manner that electronic music is both different from, but also similar to, acoustic music. The composition was still orientated to traditional musical sounds. Basis and overtones are regularly built so that different pitches can be perceived, although the sounds themselves are quite uncommon in the traditional harmonic system. Therefore 'elektronische Musik' is, indeed, music. By contrast, 'musique concrète' has a weaker connection with the traditional concept of music. The sound material included more elements of noise, and pitches in the traditional sense could not be perceived clearly. Pieces were named according to traditional musical forms, like the 'Symphonie pour un homme seul' or others, as the list of works shows.[11] Still, the sounds had their own value, their organisation told a story or made 'sense' in a more abstract way, if the listeners were willing to use their imagination.[12]

The two contradictory aesthetic concepts mentioned above did not develop separately. In the late 1950s, certain composers amalgamated them,[13] and, in 1962, even Eimert used prerecorded sounds in his 'Epitaph für Aikichi Kuboyama'. Metaphorically speaking, 'musique concrète' and 'elektronische Musik' represent two important cornerstones in a field of tension in which electro-acoustic music, now produced in a constantly growing number of studios worldwide and no longer restricted to radio stations, developed during the following years. There is a close connection between the concept of 'traditional' acoustic music and the extension of the concept of 'musical' sounds which today includes all kinds of both natural and artificial noises. Of course, the inclusion of sounds into musical material that had previously been regarded as non-musical noises is not only the achievement of electro-acoustic composers. For decades composers of acoustic music, like the futurists, Edgard Varèse, John Cage and others, also successfully extended acoustic sound material. In the sixties, even less radical young composers experimented with all kinds of noises. New percussion instruments and acoustic playing techniques were invented to produce them. The whole instrument is regarded as a sound source when, for instance, the keys of a wind instrument are pressed rhythmically, when the body of the violin serves as a drum or when instrumentalists sing, speak or shout during their performance. A new universe of sounds became part of acoustic music, which, given the varied spectrum of sounds offered by electro-acoustic instruments, seems quite small by comparison. Computer technology in particular offered so many opportunities for producing new sounds that the impression of endless sound variations occured. During the last thirty years, numerous different programs and computer music languages were developed. Today they offer sophisticated possibilities to analyze, process and/or create sounds in real-time, and in order to facilitate the work of a composer, some of them even offer graphical displays.

Very soon it became clear, however, that this was not an unmixed blessing. Composing with the computer is much more than merely organizing sounds according to the composer's ideas or imagination. It requires specialist knowledge and skills on the part of the composer, of which programming or the handling of special music programs are only two. Thus, composition with the computer is a long and complicated process. Also, the production of sound in computer music generally means that sound, an audible entity, is split into different parameters which can be controlled individually according to the program in use and to the equipment in the studio. Obviously, the terms of musical thinking have changed dramatically in computer music. Sometimes, especially when new technology is applied, there is the risk that a given program or a technology is more decisive than musical inspiration for shaping the composition.[14] Irrespective of whether this applies to electronic compositions or not, those compositions meet the aesthetic demands of the new musical material. Although producing sounds in a more extreme way, they can thus be compared with the compositions of contemporary acoustic music.

At the other end of the spectrum in the development of electro-acoustic music there are mixed forms of music for electro-acoustic and acoustic instruments. In addition to the electrical musical instruments which were built in the thirties and are sometimes still played on stage today, new technical devices were invented in the past few decades for the live-performance of electronic sounds. Also, pre-produced tapes were played on stage in an ensemble of acoustic instrument performers. Today, all kinds of synthesizers or other electronic live-instruments are common in musical performances on stage as well as tapes that accompany musicians. Whatever sounds they play: electro-acoustic instruments, sounds and music are an established part of contemporary music.

However, there seems to be a tendency for better audience acceptance when the part of amplified instruments in the score is less prominent. An example of this is Kaija Saariaho's 'Du Cristal' (1989), a piece which is quite often performed on stage. In this composition a single synthesizer is smoothly integrated into a full orchestra of traditional acoustic instruments. Pieces with stronger electronic elements or pieces for loudspeakers only seem to be more difficult to accept by certain audiences. Even today, when the loudspeaker systems on stage are much better than they used to be thirty or forty years ago, audiences are noticeably smaller in those concerts than in concerts with musicians or live-electronics and musicians. A lot has been achieved in nearly every electro-acoustic studio to produce high performance on sound systems with little or no distortion or noises and with precise impressions of space. But the audience still seems to be bored by the lack of visual attractions or is shocked and even frightened by the new and uncommon sounds they hear; often enough, especially when the lights are dimmed, people walk out during a tape performance.[15] This is quite an old problem. In addition to the individual reception situation in a radio listener's home it occurred as soon as public performances of electro-acoustic music were organized. According to Fred Prieberg, the first concerts of 'musique concrète' were given about 1950 – and failed.[16] Schaeffer therefore invented a device to move sound in space via dynamics by handling a coil in front of four loudspeakers.[17]

The distribution of sound in a (concert) space became an important field of research for the composers and technicians of 'musique concrète'. In the late 1950s, Schaeffer and his colleagues of GRM also talked of 'acousmatic' music, a music that ought to be listened to regardless of its original source. Finally, in 1974, François Bayle introduced his 'Acousmonium', an orchestra of 78 loudspeakers with various sizes. It consists of 11 different types of loudspeakers: every type has a specific sound. In order to control every pair of speakers, two mixing consoles with up to 48 output channels are an integral part of the Acousmonium.[18] The 'instruments' are sometimes even illuminated to emphasize that the loudspeakers produce the sound and not only transmit it. As with a traditional orchestra of acoustic instruments, every composer uses it in his or her own way and also differently for each individual composition. Those who have attended an acousmonium concert have surely enjoyed the lively sound it produces and have felt the aura of the illuminated loudspeakers, especially when some of those extraordinary ball-shaped ones, obviously built in the sixties, were placed among the audience. Because of the strong visual aspect and a reception situation which does not separate stage and audience, it seems that an environment like this can hardly be called a concert any more. It is rather a sound installation, one of the new genres which, by applying technological means, amalgamates different traditionally separated types of art. But there is an important difference between these new genres and the Acousmonium. The latter was designed in order to support the perception of electro-acoustic music and its visual aspects were considered to be secondary to the acoustical ones – simply its visual aspect is sometimes stronger than Bayle had planned.

In contrast to the different types of (visually supported) audio art, the new genres combine visual and acoustic art – and sometimes even dramatic art, literature or poetry – without a hierarchy between them. In art forms like this, no aspect is functional or supports the other, but all aspects are inseparably interwoven. In this respect radio art of the late 1920s and early '30s is similar to the new genres which first appeared about a generation later, at the end of the 1950s. As in radio art, the balance of the different artistic elements is made possible by the development of both

Fig. 1: The Acousmonium at the 'Audiowerkstatt Berlin', Berlin 1988. Photo: Folkmar Hein.

electronic technology and aesthetics. Due to the many different types of traditional visual arts the integration of visual and acoustical aspects also led to a larger variety of different types of new art: sound performances, sound sculptures and sound installations were invented at about the same time.

Among these three types, sound performance could be regarded as the most 'traditional', because the connection between all kinds of stage art and acoustical art is also present in opera. Helga de la Motte-Haber[19] lists the Dada performances as a direct predecessor of this type. And indeed, some of the elements, like the early integration of different artistic genres, especially the combination of sounds and language into 'sound poems', which were performed by the artist and not by an interpreter, have very much in common with the sound performances and action art of the sixties. This is not surprising, because the artists of the Fluxus movement and of Neo-Dada began to work out this new type. Most of the more musically interested artists among them studied with John Cage, at least temporarily.[20] In his own art, Cage tried to encourage the audience to be more open-minded. He advocated the unification of artistic genres and the end of separation between musical sound and every day sound as material for composition, and this, of course, also meant the inclusion of electronic sound or other kinds of electronic media into art. These elements were also part of the first sound performances. Here the use of electronic elements was limited, because in these early years of intermedia art the electronic equipment for live-performances was also limited. Sometimes enormous efforts had to be made to include it. This, for instance, was the case with the outstanding and very successful series of intermedia performances 'Nine Evenings' which were held in New York in autumn 1966. In addition to the ten artists, including Cage, and the nine engineers from Bell Laboratories, the author Billy Klüver lists a further eleven engineers and numerous technicians who were engaged in this project together with thirty-five artists, including famous painters, musicians and dancers, who were busy on stage or in the background.[21] Klüver's descriptions show the extent to which the 'different' artistic genres and the new electronic media were intricately interwoven in every performance, sometimes involving the audience as well. The spectators were asked to leave their seats and walk around, and in 'Physical Things' by Steve Paxton (artist) and Dick Wolff (engineer) they had to walk under a net of wires inside a plastic tent with a small electronic device in their hands consisting of amplifier, loudspeaker and a magnetic system. It enabled the visitors to listen to the otherwise unaudible sounds that were 'in' the wires. Because each wire was connected to a different tape, every listener could assemble her or his own concert by moving individually inside the tent. Additionally, in a second tent there were slide – and live – shows, so that even a mixture of sound performance and sound installation was presented.

Compared to the sound performances of today, this event of 1966 was spectacular. Additionally, its underlying concept of an intermedia art performance reached beyond its time and became a model which might be called 'classical' today, because it was adopted by many artists, using a wider range of electronic equipment. Hence nowadays, unless the artists become engineers themselves, individual technical development for the arts is less needed and rarely done. The engineering work of the 'STEIM' in Amsterdam may therefore be called exceptional. Specializing in electronic live-art, 'STEIM' is a center for performance artists from all over the world who receive help with modifications of their technical equipment and who, in return, often initiate alterations of the equipment developed there. Much has been done there in the past few years in the real time processing of visual data (BigEye) or in real time soundprocessing (LISA).[22] Laetetia Sonami for instance, an artist who stayed at 'STEIM' several times, works extensively with real time sound processing during her performances with the 'Lady's Glove',[23] a small control device connected via MIDI with a sampler or other electronic equipment. The control contacts and cables are placed inside a lady's evening glove. Sonami switches the sampled sounds by a small movement of her hands and fingers and processes them with another small movement or the control of the distance between her hands and arms. It is a very elaborate instrument which she plays virtuoso-like on stage. During her concentrated performances it looks as if she is taking sounds, speech and music out of the air. Despite the delicate interplay of literature, noises and sound, it is difficult to decide if this is still a performance or rather a concert in the traditional sense.

Like the performances, the art of sound sculptures has a long history of its own, too. Sounding figures or sculptures like 'Aeolsharfe' or musical clocks and machines have a long tradition. Sometimes early sound sculptures can be regarded as genuine and elaborate pieces of art, not only as craft devices. An unusual example is 'Tipoo's Tiger', a sounding machine manufactured about 1795 by a French artist for Sultan Tipoo in South India. Its sounds are noises and there is also a political dimension to it. The 1,75 m wooden sculpture shows a tiger lying on a white man; the snarls and roars of the beast as well as the cries of its victim are audible. This sound sculpture was referred to directly by Stephan von Huene in 1967 when he built his first audio - cinetic object 'Kaleidophonic Dog', which has an electric drive and is made from wood, leather and parts of a xylophone, percussion and organ pipes.[24] Von Huene often uses these materials and sometimes works with the transformation of language into sound. 'Zauberflöte' for instance, a group of four sound sculptures made out of metallophon/xylophon plates, organ pipes, lamps and technical devices in a wooden box, is based upon Schikanaeder's libretto of Mozart's famous opera. Von Huene transforms the complete text into sound. According to their phonetic appearance he allocates certain sounds of the vocals to the libretto. In his arrangement, the organ pipes of the type 'vox humana' stand for the singers' voices and the other organ pipes and plates represent the instruments of the orchestra. In order to guide the viewers, von Huene transforms all verbs of the libretto which express sensual attractions into the visual attraction of lamps on the four objects by using the theory of neurolinguistic programming.[25]

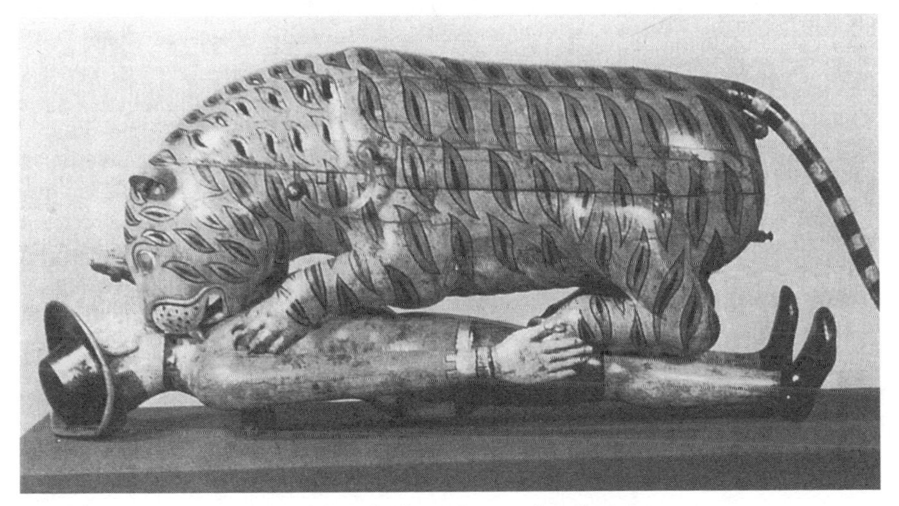

Fig. 2: Tipoo's Tiger c 1795. Victoria and Albert Museum, London.

What is striking here is the reception conflict between the visual and the sound elements which arises because a traditional piece of performed art is used in a sound sculpture. During the opening hours of an exhibition of 'Die Zauberflöte' by von Huene there is a constant admission of visitors while the sound- and light part runs continuously without clearly marked breaks between the end and a new beginning.

Hence the visitors are hardly able to attend von Huene's 'Die Zauberflöte' in the traditional way of an opera – completely from the beginning to the very end and in a group of persons – but individually in the reception tradition of sculptures.

This reception conflict, of course, comes up with every piece of intermedia art which both changes its appearance in time and has visual aspects that can be received individually. In many cases it will immediately become clear to the visitor that a 'complete' reception of a work is impossible. Although sound and music in the traditional sense are an important part of this new art form, it seems as if the issue of a complete work of art cannot be answered in the tradition of music, but in the tradition of visual art. This will be illustrated by dealing with two sound sculptures produced in the last few years.

'Cut Pipe' (1992) by Gary Hill is a long metal pipe cut into two pieces. Two small round monitors are fixed to the ends at the centre of the sculpture. On the screens, two loudspeaker membranes are shown which are manipulated by two hands. While touching and squeezing the membranes, the visitors hear the noises these hands are making. Even this short description makes clear that this sound sculpture could not exist without sound. But in order to enjoy and understand this piece of art the visitors do not need to see and hear all the variations shown on the screen. The other example, 'Whirlwind' (1996) by Laurie Anderson, is even more extreme. The sounds are processed songs and stories about time; they whirl around in a circle marked by light. There is only sound and light, a sound sculpture built by two immaterial elements. This means that the sound sculpture only exists when it is 'switched on'. Again, visitors do not need to listen to all the sounds in order to understand the concept of this most 'unsculptural' sound sculpture which can only exist by means of electricity.[26]

Whereas sound sculptures are defined by their own concrete figure, sound installations artistically describe a given space or room visually and acoustically. Because installation art was invented in this century as part of visual arts, sound installations seem to be the most radical form among the three new types of art. To be sure, earlier forms of sound installations exist. The space-including church compositions of the Italian *cinquecento*, for instance, are examples of the acoustical component of sound installations. An even better example is Satie's 'Musique d'Ameublement' mentioned earlier. For the visual component, special interior architecture may be regarded as a precursor. But unlike the two new types of art described above, intermedia performance and sound sculpture, there is no preceding art combining both elements, the acoustical and the visual. Sound installations, therefore, represent a completely new type of art.

Having no direct predecessors has perhaps made its acceptance easier. The new art type of 'sound installations' with its amalgamation of the aesthetics of music and art as well as the growing role of electronic media seems, at least at present, to demand less aesthetic justification than the other new types of art which have direct precursors: the distance to the traditional art forms and their aesthetic foundations is simply too large.

There are two different ways to describe 'space' with sound installations: it may either be shaped or created. In 1967, Marianne Amacher transferred sound from different places into her sound installation 'City Links', thereby shaping the space of this installation,[27] whereas in 1967/68 Max Neuhaus created a space with his sound installation 'Drive-in Music', when he installed short range radio transmitters in trees along a boulevard to be received on the radio of the cars passing by.[28] With this installation Neuhaus removed a space that was 'sounding' from its natural surrounding and marked new boundaries with it.

Here, as in every sound installation, the audience's role is different from that in other forms of art. Because of the reception situation the audience is inside the installation and sometimes even becomes part of the work of art. This is especially the case in 'interactive' installations, a term introduced by Peter Weibel in 1989[29] to describe art which reacts to the behaviour of its visitors. Interactive sound installations had already existed at that time, although the name was not there. In 1985, for instance, John Driscoll introduced 'A Hall is All',[30] a sound installation that both shapes the room and reacts to the presence of visitors: in the room a few well shaped loudspeakers and rotating horns emit sounds that vary with the number of visitors present. These sounds are received by microphones and processed electronically. Hence the sound changes in accordance with the number of visitors. This means that the shaping of the room varies in four dimensions, in its three dimensional space as well as over time: visually with the different number of people present and acoustically with the different sounds.

Reactive or interactive sound installations are very popular today. Meanwhile, in addition to the public participation they offer other features of visitor involvement. Sometimes the artistic intention of sound installations seems to be the evocation of strong or authentic feelings. For instance, 'Intersection', a sound installation presented by Don Ritter in 1993, is a reactive installation in a completely dark room. Guided by lines, the visitors cross an acoustically imitated highway; cars stop with squeaking breaks or crash into each other if a visitor 'stands in their way'. The effect on the visitors can be embarrassing, especially when many people are present in the installation and also bump into each other.[31] A much more careful handling of its audience is given by the sound installation 'Wege in die Stille' (Ways into Silence) by Götz Lemberg (1996),[32] offering a comforting experience to its visitors. It consists of five different small rooms. In one of them, lengths of light material are hanging from the ceiling. While crossing the room, the visitors are touched softly and hear different sounds close to their ears. In another room, the visitors have to walk barefoot on large pebbles, producing a soft sound. The installation is illuminated carefully in the last room, long periods of silence alternate with low noises from tape which are similar to the residual noise produced in the ear itself after a very loud and busy day. So, electric technology is still necessary, although the artist sometimes produces sound installations with no audible sounds at all.[33]

Audience consideration is also important for Robin Minard. He explicitly speaks of the social responsibility he has as an artist of public sound installations in shaping the sounds of everyday life.[34] Therefore he wants his work to be accessible and interesting for everybody. In his 'Still / Life' series, for instance, he works with many small high-frequency loudspeakers which are arranged in an area. Because of the soft and whirring sounds of the loudspeakers, the installation has a lively atmosphere. The shape of these areas corresponds to the architecture of the room in which the installation takes place. In the crypt of the lower church in the monastry of 'Unserer Lieben Frauen' in Magdeburg, which dates from the eleventh century, Minard arranged 2300 loudspeakers in a narrow closed area which then looked like a cloud in relation to the unevenness of the floor. In a modern room, as in the gallery of the DAAD in Berlin,[35] he arranged a geometric figure in the shape of the shadows from a room partition. In both cases, these sound installations create a unity of sound and space. They are unique, and therefore, in more than one respect, fulfill the traditional conditions of a work of art, the concept of 'l'art pour l'art' which exists independent of an audience.

Fig. 3: Robin Minard: 'Still / Life'. Installation with 2300 high-frequency piezo loudspeakers. Kloster Unserer Lieben Frauen, Magdeburg, May-June, 1996. Photo: Hans-Wulf Kunze.

One can observe the traditional aesthetics of the nineteenth century in many other contemporary works of the new electricity-induced genres, for instance in all of the contemporary and most of the earlier works dealt with above. This neither means that all contemporary art refers to the aesthetics of a bygone time nor that works of art with some observable connection with traditional aesthetics are necessarily conservative and backward-orientated. Art represented in the new genres is so completely different from that of the past that a comparison can only refer to certain aspects, not to art in general.

In the new genres dealt with above, a change in the concept of music can be observed. This is particularly true of sound sculptures and sound installations which include strong visual elements. This refers not only to the extension of the sound material and the less directed attitude towards time (an aspect of that originated with some of John Cage's compositions and with minimal music), but a specific type of musical concept was developed by these genres. In a sound installation like Don Ritter's, the term 'composition' no longer seems to be appropriate for the sound component of the installation, although it is carefully organized. How can art like this be analysed, especially as it works partly without a visual aspect (the room is darkened completely) and with an acoustical aspect that does not fit any conventional musical form? Similar questions have to be asked regarding the other examples as well. If, for instance, only the sound aspect of Robin Minard's 'Still / Life' series is examined, the extremely slow progress in time is remarkable, but this impression never occurs when examining the sound installation as a whole. Likewise, simply listening to the sounds of Gary Hill's 'Cut Pipe' would make no sense at all. In John Driscoll's 'A Hall is All' one cannot listen to the sound separately, because it is produced and shaped by the interaction of the visual electronic equipment, by the audience and by the room in which the equipment is installed. The intermingling of traditional artistic forms in the new genres is so complete that an analysis or description of a given piece of this art would make no sense if subdivided into its traditional elements, for the interaction of these elements in the piece of art could not be observed if analyzed separately. Although perceived on different levels and with different organs (ear and eye), they are nevertheless perceived and understood as complex units or 'super signs', as these perception units were called in the information theory and semiotics of the 1960s. Further research is needed in order to establish appropriate tools for the analysis of the new art. Nevertheless, it is obvious that today's art of the new genres is different from the one produced a century ago.

Aspects of the traditional aesthetics are present on different levels. The majority of pieces of the new genres seem to represent closed forms, which is one reason why they can be called 'works of art' in the traditional sense. This, of course, refers to the complete work. Even if the sound aspect (or another one), as in most of the examples above, shows no tendency towards 'closed forms', there may be other aspects, for instance the work's three dimensional extension, which can be called 'closed'. Therefore, a sound sculpture or a sound installation can be called a closed form, if only the sound aspect defines the space of the work against an unspecific environment. The sound aspect itself might be rather static and exhibit no form in a musical sense.

Yet another aspect of the new genres refers to traditional art and its aesthetics: it is the 'aura', a term which describes the distance of a work of art from the audience and which insinuates some sort of magic. 'Aura' occurs due to its uniqueness and cannot be reproduced by technology. This, at least, is the definition given by Walter Benjamin in his famous essay 'The Work of Art in the Age of Mechanical Reproduction',[36] first published in 1936. Here Benjamin states that the ability of technical reproduction destroys the 'aura' of a given work of art because of the many copies that might exist besides the original. Benjamin seems to have welcomed the loss of 'aura' in art, because it also meant the loss of its autonomy in favour of a new sociological function in mass reception. As far as art of the new genres is concerned, Benjamin was partly right. There is an 'aura' in many of the works of the new genres which cannot be reproduced completely, because even a video recording can only document parts of the work, especially when it involves three-dimensional space. This statement is supported by the observation that no recording of the contemporary art of Laetetia Sonami is available. The 'aura' of her public performances which some people call 'magic',[37] will probably be lost when played too often in a private environment. However, the question remains whether this kind of contemporary auratic art can be called 'autonomous'. The vital role of audience participation in some of the works contradicts this idea. It rather confirms another of Benjamin's theses about the relationship between art and its audience, namely the direct connection between simultaneously experiencing and reviewing art. The importance of art for society, he states, is recognizable in the coincidence of both a critical and a pleasurable attitude on the part of the audience.[38] In some of the new works of art a new aesthetic approach is visible which is in line with Benjamin's ideas. On this basis it seems possible to build on older artistic traditions while starting something new at the same time. Further developments in this field will be interesting to watch.

Notes

1 F. Trautwein, 'Bedeutung und Wesen der elektrischen Musik', *Jahresberichte der Hochschule für Musik Berlin-Charlottenburg*, 1929/30, 31.

2 'Nur ein gewissenhaftes und langes Experimentieren, eine fortgesetzte Erziehung der Ohren, werden dieses ungewohnte Material einer heranwachsenden Generation und der Kunst gefügig machen.' F. Busoni, *Entwurf einer neuen Ästhetik der Tonkunst* (neue Ausgabe von H. H. Stuckenschmidt, 1953, first edition, 1907), 42.

3 It does not matter that Satie was completely misunderstood by the audience of his time. At the first public performance of 'Musique d'ameublement' in a Paris gallery in 1920, contrary to the wish of the composer, the audience stopped talking, sat down and listened to the two piano players (Erik Satie and Darius Milhaud) as in a concert. See: H. de la Motte-Haber, *Musik und bildende Kunst* (Laaber, 1990), 268f.

4 The first two terms appear even on the same page of *Der deutsche Rundfunk*, 1929, Nr. 49, 1549 and in a second article of the same volume (1551) the third term is used.

5 F. Stiemer, *Der deutsche Rundfunk*, 1930, Nr. 17, 65, translated by the author.

6 W. G., *Der deutsche Rundfunk*, Heft 21, 10.

7 Perhaps the best translation is 'applied art'.

8 *Der deutsche Rundfunk*, Nr. 22, 30 May 1930, 10.

9 This was at least the setting in public radio art 'concerts' in the 'Rundfunkversuchsstelle' in Berlin or at the first radio music festival in Baden-Baden 1929.

10 Situation de la Recherche, *Cahier d'Etude de Radio-Television* (Flammarion, 1960), Nr. 27/28, 5.

11 In: 'Recherche musicale au GRM', *Revue musicale*, 1986, Nr. 394-397, 258ff. Most of the early works have titles like 'Concert', 'Etude' or parts are named 'Andante', 'Allegro', etc.

12 Some compositions of the 'musique concrète' that were produced in Paris at the GRM can be called acoustical documentations, too.

13 Stockhausen was the first to do so by composing 'Der Gesang der Jünglinge' in 1956/57.

14 Experimental music is, of course, interested in the sounding result of the experiment. In experimental computer music, very often mathematical formulas serve as construction rules, regardless of the music they produce. Among the many examples, applications of fractals, formulas that produce self similarities on micro and macro levels, seem to be most illustrative. Their visual applications are famous because of the beautiful patterns they leave on the monitor, whereas composed fractals only seldom seem to produce no audible self similarities at all.

15 This problem has never really been studied in detail but has often been observed by the author (i.e. during a concert on a congress about New Music in Darmstadt, 1997). It is, of course, a point of discussion among composers, technicians and other people who are interested in electro-acoustic music.

16 F. Prieberg, *Musica ex machina* (Berlin, Frankfurt, Wien, 1960). This book is a long essay and a lively description of early electro-acoustic music and its forerunners rather than a scholary text. Therefore, the date of the first public concert is only a rough reconstruction from other dates and statements in the text.

17 Prieberg, op. cit. (note 16), 127ff. Prieberg describes a concert that took place in Basel 1955. A picture of Schaeffer with this 'console', taken in 1952, is printed in Recherche musicale au GRM, op. cit. (note 11), 144-145.

18 F. Bayle, 'A Propos de l'Acousmonium' in Recherche musicale au GRM, op. cit. (note 11), 144-146.

19 H. de la Motte-Haber, 'Klangkunst – eine neue Gattung?', *Klangkunst-Katalog* (München, 1996), 14.

20 K. Thomas, *Bis heute: Stilgeschichte der bildenden Kunst im 20. Jh.*, (Köln, 9th ed. 1994), 242ff.

21 B. Klüver, '9 Evenings: Theatre and Engineering; über die Benutzung des Mediums Ton durch Künstler', *Für Augen und Ohren, Ausstellungskatalog* (Berlin, 1980), 89ff. Although Klüver called this 'Theatre and Engineering', his descriptions make clear that this term does not do justice to these events.

22 STEIM: *De zoetgevooisde Bliksem*, (Catalogue, 1993) and grey material.

23 A picture and further information on 'Laetetia Sonami' see *Klangkunst-Katalog*, op. cit. (note 19), 138f.

24 K. Schmidt, 'Eine Einführung in das Werk von Stephan v. Huene' in *Stephan v. Huene, Klangskulpturen. Ausstellungskatalog d. Staatlichen Kunsthalle Baden-Baden* (Catalogue, Baden-Baden, 1983), 11ff. (for Tipoo's Tiger see the footnote, 23f).

25 S. v. Huene, 'Notizen zur "Zauberflöte" ', *Musik und Sprache, Sprache der Künste III*, Katalog der Inventionen 1986 (Berlin, 1986), 100ff.

26 Both sound sculptures are described in more detail in: 'Laurie Anderson' (34f.) and 'Gary Hill' (70f.) in the *Klangkunst-Katalog*, op. cit. (note 19).

27 *Klangkunst-Katalog*, op. cit. (note 19), 285 (Chronik).

28 A description of this sound installation can be found in: G. Föllmer, 'Töne für die Straße', *Klangkunst-Katalog*, op. cit. (note 19), 216ff.

29 P. Weibel, 'Momente der Interaktivität', *Kunstforum International*, 1989, Nr. 103/4 Sept./Oct. and P. Weibel, 'Für eine interaktive Kunst' in *Ars Electronica 1989* (1989), 87ff.

30 *Inventionen* (Catalogue, Berlin, 1986), 134f.

31 This installation belonged to the 'Sonambiente' – Festival, too, and is described in the *Klangkunst-Katalog*, op. cit. (note 19), 124f.

32 *Klangkunst-Katalog*, op. cit. (note 19), 90f.

33 Like in a church in Berlin-Neukölln, 1997.

34 In an interview with the author in September 1997. Aspects of this also can be found in R. Minard, *Sound Environment: Music for Public Spaces* (Berlin, 1993).

35 See: R. Minard, *Klanginstallationen, Künstlerhaus Schloß Wiepersdorf* (Künstlerhaus Schloß Wiepersdorf, 1996).

36 The English title is 'The Work of Art in the Age of Mechanical Reproduction'.

37 Some members of the audience, who attended a performance of Laetetia Sonami in Berlin in August 1996, made remarks to this effect.

38 W. Benjamin, *Das Kunstwerk im Zeitalter seiner technischen Reproduzierbarkeit* (Frankfurt/M., 1976), 37. In the German original Benjamin wrote: 'Dabei ist das fortschrittliche Verhalten dadurch gekennzeichnet, daß die Lust am Schauen und am Erleben in ihm eine unmittelbare und innige Verbindung mit der Haltung des fachmännischen Beurteilers eingeht. Solche Verbindung ist ein wichtiges gesellschaftliches Indizium. Je mehr nämlich die gesellschaftliche Bedeutung einer Kunst sich vermindert, desto mehr fallen die kritische und die genießende Haltung im Publikum auseinander.'

Bernd Enders

MUSICAL EDUCATION AND THE NEW MEDIA: THE CURRENT SITUATION AND PERSPECTIVES FOR THE FUTURE[1]

1. *Media technology at its current stage of development*

> The scenario leaves almost nothing to be desired.
> On the large screen in your living room, you can download movies and news on demand. With the remote in your hand, you do not have to leave your couch to go shopping, read the newspaper, learn or play. The teleworker of the future has his office in the next room. Free from all constraints brought about by rigid office schedules or traffic problems, he sends the results of his electronic activity directly to his employer or customers. Using his computer and his video telephone, he collaborates with experts around the world; distance doesn't matter to him anymore. During his breaks, he reads his email.

This prognosis is not meant as a joke or a parody. Uwe Jean Heuser uses it to introduce his analysis of a 'brave' or 'ugly new world' transformed by media technology – a world that our modern information society is bringing forth.[2] He continues:

> 'Few technologies have given rise to such great hopes and at the same time to such dramatic fears as the info highway, the symbol of the information society. Some of us mainly set their stakes in the technical possibilities offered by tomorrow's interlinked world. All across the globe, people can communicate with each other, form electronic communities or download any kind of imaginable information. Others foresee evergrowing floods of trivial entertainment; they are frightened by visions of people isolating themselves at home, sitting at their monitors and withdrawing totally from reality while losing themselves in so-called cyberspace.'

The French philosopher and sociologist Jean Baudrillard, for one, is convinced that the virtual man who squats motionless in front of his computer, makes love through the monitor, gives lectures via teleconference – who is, in other words, restricted in his mobility, could also become mentally disabled. Paul Virilio, for his part, fears that home interiors could become so deconstructed that they eventually might lack an entrance door or any kind of other opening. Botho Strauss sees an absolute totalitarianism looming up ahead in the near future, a system that does not need to make any heads roll because it makes the heads themselves superfluous. Another ironic variation of this attitude is, 'New technologies are always offering new solutions to problems that had not existed before'.

In Germany we have, among others, Hartmut von Hentig, the well-known educationalist and director of the North Rhine Westphalia Experimental School (*Laborschule*). He represents a worst case attitude typical in the German educational system, where computers are disparaged as posing a fundamental obstacle to any attempt at teaching. 'Computers keep children in their seats, limit their movements to the area between the monitor and the keyboard, neutralise their senses, break off their contacts with the outside world and limit their thinking to the programme's question-and-answer schema. In principle, the use of computers obliterates all of the efforts which have been made in pedagogy since the beginning of our century.'[3]

Nicholas Negroponte, the founder of MIT Media Lab, holds such attitudes to be fundamentally wrong – children have to learn to deal with computers anyhow. Moreover, he can refute each and every one of von Hentig's statements; by sharing their experiences at the computer with each other, children make new friends, have fun while learning and experience new qualities of life. In his recent book 'Being Digital', a US bestseller, he claims that a new world and a new lifestyle are beginning to emerge.

This new world is characterised by everything related to the production, distribution and use of information, especially of information provided by computers and by information highways. He presumes that we will soon have 15 000 TV channels to choose from, that we will be able to view several million books online and download billions of individual newspaper pages. 'Libraries', in the etymological sense, become extinct; they are transformed into multimedia data archives.

The internet, the network of networks that was originally kept up mainly by American universities, has already established itself as the precursor of the information highway, interlinking computers from around the globe. Schools are discovering the new technology as a basis for internationally available and exchangeable school magazines or as an efficient tool for both teaching and learning. Many companies are starting to employ their first teleworkers. Other catchwords are: video on demand, teleshopping, telebanking, video conferencing, netsurfing, channel chatting, etc. Virtual museums are in the making, a hopelessly large amount of multimedia products with educational subjects already exists, although high quality is more the exception than the rule. Even multimedia higher education is already being discussed as a possibility.

There is no more doubt about the fact that information highways will transform our societies in fundamental ways. It is high time that we considered, in detail, all of these new technologies' positive and negative effects that we can possibly foresee at this point.

The rapidly emerging information society will demand great flexibility from people as well as from political systems. The Gutenberg galaxy, i.e. the era of books, required several centuries in order to expand to its full size; radio and television still took a few decades until they reached full technical development. But the newest technological innovations are succeeding each other more rapidly than we can record them or react to them. Thus, the usual one-generation cultural lag in educational institutions will have ever more devastating consequences. Social change is already under way. Information has become our society's most important commodity; digital systems store and transmit information in almost any imaginable form. Modern computer technology takes over many of the data management chores that used to be reserved for human intelligence.

tic
of th

2. *Determ*

2.1 The technologic
Until now onscreen
'new media'[6]: electronic
transmission, digital FM ra
– the latter only until recently,
CD-ROMs.
Essential current technological inn
computer systems, which are instigati
cation of media and communication te
separate areas.

2.1.1 Digital media technology (multimedia)
The catchword 'multimedia', most of all a publicity
industry's sales, also signals an essential transformatio
first, the computer served scientists mainly as a fast calcu
fice employees as an efficient data archive and as a flexible
since it started being used as a game machine, as a MIDI w
arrangements and as a graphics system, it has metamorphosed
cultural contraption that holds a useful purpose for anyone. The
these developments have reached is, for the time being, multimedia te
computer based media technology, which means digital processing of all
formation and perception (limited for the moment to seeing and hearing, ex
later on to feeling and touching).

The standardisation of storage media is already well under way. The CDPlus
successor to the conventional audio CD, but at the same time also supersedes the extremely successful CD-ROM, and, of course, the audio cassette. Soon it may even re-

ptical) storage media such as photos, film

ed particularly to musical subjects. The fa-
oduced CD-ROMs dealing with the systemat-
ven's 9th, Mozart's Dissonance Quartet or
itorial decisions were not fostered so much by
t rather with the goal in mind of demonstrating
a both *eye*- and *ear*catching way.
s pop star video clips, but also catering to children's
he scene and probably made the most money. Con-
ly elaborate and in many ways exemplary CD-ROM
wn as Prince. Furthermore, there was also a simple
the Beatles, taken from the film 'A Hard Day's Night', as
r Gabriel, David Bowie and others. Later on, similar CD-
s and rock groups were released (some of them of incredi-
least one positive example can be mentioned: 'Das Auge

OM bestsellers can now be found in other categories than be-
e computer freaks who used to watch and listen to Stravinsky's
as multimedially packaged, but it is now normal PC users who
e market. In other words, current sales statistics are dominated by
ries, encyclopaedias, dictionaries and erotica in the most diverse
interaction quality levels imaginable. CD-ROMs with content re-
n and history also receive wide attention.
ltimedia business, music plays an important role, sometimes in unex-
s such as the Baedeker tourist guide to the US, in which you can hear
John Lee Hooker.
side these developments there are also numerous music programs featur-
sequencing, MIDI composing, computer-aided musical instruction, music
ample editing and arranging, etc., that are not only intended for pop profes-
but for amateurs as well.

Global networking (information highways)

s expected that the world's information and communication systems will merge
gether, not only via the internet. Digital radio and TV stations are to be equipped
ith feedback channels in order to be able to take the audience's reactions into account, for instance its musical tastes. But maybe someday only the latest news, along
with soccer games and shows with direct viewer/listener participation, will be transmitted live, while all other films and music will be downloadable on demand.

Already Bertolt Brecht, while writing his radio play about Lindbergh's Atlantic
crossing, had the vision of radio as a bi-directional communication medium. By
using a feedback channel, so he imagined, the listener should not only be able to receive, but also to transmit.

Once video conferences can be held in real time, playing music together as well as composing and arranging together should also become possible with a video link between the collaborators.

Teleshopping, the telephone, the fax machine, online services of all kinds, digital radio with integrated visual information that can be watched (for instance frozen shots of the conductor, the orchestra or the score), CDs, videotapes, films and photo CDs – everything merges together with the computer as a central control unit, serving as a terminal outlet for the net and as a multifunctional interface between the user and whatever universe of data he would like to access.

The only ultimate difference between a multimedia PC and a digital TV, apart from the different transmission channels (telephone lines and radio waves), is the question whether the user is sitting alone directly in front of a screen or sitting at some distance from it, possibly in the company of other viewers. In the end, it is solely a socio-dynamic decision opting for either individual or collective information processing, in the form of one-dimensional reception or two-dimensional interaction.

The music forums on the internet are already offering quite a variety of information – multimedia World Wide Web pages on Bob Marley, Peter Gabriel, Yello, but also musical research forums with subjects ranging from the baroque period to computer music. It is already common practice to exchange complete musical pieces (usually amateur arrangements of well-known pop songs) in the form of MIDI files or sound files.

It would already be possible to have, say, piano lessons take place over the internet by sending the interpretation of a piece by the pupil or by the teacher (who might be on concert tour) as a MIDI file for the diskette piano; the lesson could also take place with live video contact.

2.1.3 Virtual reality (cyberspace, cyberworld)

It is hard to assess the newest technological development, virtual reality, since it is still in its early stages. For starters, virtual cities and meeting places stylised as cafés can already be found on the diverse networks; there are shops on the internet as well as communicative games and virtual museums with a spatial effect when stereoscopic viewing machines are used.

The sense of touch is the only element lacking in order to be able to offer a complete virtual scenario with a total physical effect. However, it is by no means certain whether this final feature could not be incorporated into virtual reality, so that, in the long run, real and virtual experience would become almost interchangeable. Cybersex controllers with tactile sensors have received wide media coverage and represent the first attempt to overcome this final barrier.

From the perspective of control technology, the next step toward virtual music-making would then be relatively short. Data gloves and other customised control devices have been used in the past years in concert performances (Michel Waisvisz,

Laurie Anderson). From a technical point of view, manual control mechanisms are the most antiquated link in any human-machine interaction. The alphanumeric keyboard is, so to speak, a relict of the Gutenberg galaxy.

Actually, before its current use for manual actions in the virtual world, the invention of the data glove as a universal control device for digital processes was originally intended for musical purposes – in order to produce MIDI data for musical performance.

To summarise: in the near future, there will be no ultimate technical differentiation between media technology, computer and communication technology. All areas of application which were hitherto partially separated due to technical constraints will all eventually operate solely on the basis of digital data processing and, in the end, will thus converge due to this total compatibility.

2.2 The anthropological perspective

Digital media have a threefold effect on the concrete actions of people who have a computer with access to the Net:

- new modes of *human-machine interaction*
- new modes of communication due to global networking (*global village*)
- new modes of modelling and simulation (*virtual reality*)

2.2.1 Interaction

It is not just people anymore who interact with each other, it is also machines with machines and people with machines. Technical evolution began with tools that extended, strengthened or eased the strain on human body parts. Then came the independent machine running on an exterior energy source, and finally the automaton that could accomplish complete tasks without needing any further help. Computers are perfected automatons, since they take on not only physical but also mental human taskloads, and thus possess the ability to communicate to a certain extent.

At Stanford, work is already under way on an artificially intelligent musical instrument that provides a musically meaningful accompaniment, grasps the musician's mood, and maybe even asks the musician, at the start of a session, what is to be played. Pure science fiction? Maybe so. But a flexible automatic musical accompanist already exists.[8]

It can be foreseen that, very soon, access to world knowledge and the acquisition of know-how and skills via computer will be one of everyone's daily activities in industrialised societies. Interactive learning programmes that can be produced by means of special author systems belong to the most important and best-known multimedia programmes on the market. This will be a special testing ground for the intelligent organisation of interaction and communication with machines. Teaching/learning at the screen will depend essentially on the degree of flexibility in which programmes are able to adapt to the users' performance, learning progress, mistakes and difficulties in understanding.

It is estimated that it takes five to ten times more effort and work hours to write a good teaching/learning programme than to produce a quality schoolbook. It is still relatively expensive to acquire license permission for the use of pictures, music examples and scores. These are some of the reasons why the available quantity of good didactic software is still quite small.

The entertainment industry (i.e. game, film and video companies) has the best conditions and financial assets for becoming the leader in this segment of the market. 'Edutainment' is the catchword which indicates that whoever is able to offer attractive or even seductive products will be setting the pedagogical course for the future.[9]

2.2.2 Communication

Computers and digital networks not only offer access to world-wide information, but also make total communication between individuals and groups possible, regardless of timezone or place.

On the one hand, we will have communication that is totally determined by the media; on the other hand, there will be communication that is media-selective. We connect our senses linking them through media that change or limit information content.

Knowledge telling us which information is necessary and which is superfluous will become centrally important. Customised browsers, specially tailored pathfinders, helping to navigate through endless mountains of information, will become essential. Netsurfing will only make sense with the help of artificially intelligent agents that help to wade through the data morass, and that operate using specifically directed selection criteria. Hopefully, such navigators will have an 'eye' for what is essential and will be able to detect and screen out any manipulation caused by third parties or special interest groups. Neil Postman: 'The open flow of information on the data highways hardly lets itself be examined with any critical competence. Pure nonsense is given the same space as serious websites. In order to find one's way in this jumble of information, a good education is needed, and this is not provided by the data networks themselves.'[10]

Some educators already fear that Internet providers and edutainment CD-ROMs will have a 'deschooling' effect that will gradually make schools lose their *raison-d'être* for society. This fear is not totally unwarranted. Bear in mind that politicians keep a close watch on schools, considering them to be expensive and not particularly efficient. The American author Lewis Perelman is even of the opinion that the new hypermedia will make schools totally superfluous. The first projects using this approach deal with the possibilities offered by virtual classrooms, or try out linking schools on different continents or teachers with their pupils sitting at home in front of the screen.[11]

2.2.3 Experiencing simulated realities

Digital processing of visual and auditive phenomena permits the creation of virtual universes, the attempt to perfectly simulate realities and nonrealities, the construction of novel worlds of experience, dreams and fantasy. For many people, the differences between the real world and fictional worlds, already blurred by television, will become even more indistinct.

Virtual reality seeks to create the perfect illusion; film and television images as we know them will seem primitive in comparison. It will even be possible to pay a virtual visit to a museum of musical instruments somewhere in the world, to stroll through a library, a recording studio, an opera house. Environmentally unsound, time-consuming people mobility will be replaced by speedy virtuality that can download information and offer entertaining experiences without wasting the world's resources. Instead of physical experience there will be a kind of digital headiness; real physical experience will be superseded by the sensuality of electronic images and sounds (later on, the transmission of taste and smell information might also become possible).

As a consequence of media omnipresence, the possibilities of confronting the real world in a primary, hands-on experience will tend to diminish as a consequence of media omnipresence. (Many German children are already convinced that cows are normally purple, since they only know them from advertisements for a certain brand of chocolate.)

3. Technologically Induced Transformations of Cultural Life: Five Theses

Let us dare to formulate some theses concerning the changes in social and cultural circumstances that could result from the technological innovations mentioned above:

3.1 First Thesis: Digitalisation modifies the information itself

From a technical point of view, digital technique loses certain fine nuances in the process of quantification (for instance those fine nuances made by a musician which the listener recognises as having artistic value, but which could be lost on a disc piano). Since minute elements that are nevertheless important in the creative process are filtered out by digitalisation, it is up to artists and technicians to become aware of such fine nuances of timbre and rhythm modulation in order to make sure that digital technology takes them into account. There is no reason why this should not be feasible.

Digitalisation has to be so finemeshed that the recipient cannot hear the difference between the digitised result and an analogue sound or musical structure. Every MIDI arranger has noticed by now that there is a great difference between the quantification of music in a printed score and the quantification applied in sound control data. Modern software is able to do justice to that difference. The loss of sound information in the reduced data on a minidisc is obvious even to an ear that has not been trained in hi-fi subtleties.

No matter whether the technical reduction of data is due to deliberate manipulation, lack of knowledge or of scruples, the problems it causes still produce aesthetical realities notwithstanding. New computer software programmes (for instance, *Band In A Box* or *Circle Elements*) define pattern-oriented musical structures or looplike fragments of sound. Mechanically exact repetitions of sound layers such as those designed with the help of a sequencer create new hearing habits that open the door to unforeseen aesthetical dimensions. Educational institutions must be ready to respond to such new habits and styles.

Historical steps such as the invention of keyboard instruments, the adoption of well-tempered tuning and, finally, that of the MIDI system (which uses well-tempered tuning as its basis) are all based on the reduction of musical information. These reductions were, in effect, technically induced and they unconsciously helped certain musical structures to prevail. Thus, they contributed to the consolidation of certain aesthetic norms.

Computers must not necessarily reinforce the habit of thinking in fixed patterns (Neil Postman warns strongly against this[12]), but they very often do so in practice. If interaction processes are to have a minimum degree of flexibility, they demand a high quantity of programming and computation. Unfortunately, though, software tends to be produced as economically as possible. In many cases the necessity of avoiding fixed procedures for certain tasks is not foreseen at all. Learning programmes, for instance, already define certain partial aspects in advance (such as the degree of quantification in musical performance tasks), and only those problems or solutions which have been foreseen become part of the learning process.

All those things which lie far apart from each other in real time and space are brought together by multimedia and data networks until they are juxtaposed as icons on a desktop or in the form of internet addresses waiting for a mouse click. It can hardly be foreseen just how our current understanding of knowledge, experiences, social and cultural relations and historical processes will change in the wake of the planned digitalisation of all purportedly relevant information. The only thing that is certain is that our scientific, cultural and philosophical models of the world and of human beings will have to address this question.

3.2 Second Thesis: Multimedia modifies the value and the hierarchy of the arts

New ways of linking video and audio are being invented: new methods of synchronisation such as granular synthesis, impressive performances such as those of Bob Ostertag, or even video clips and video art which are quite often completely computer generated.

The video clip is the forerunner in the new amalgamation of the arts. Once video is digitally broadcasted, transmitted and exchanged, it will become especially easy to work, using only one standard of data processing, since there will be no more need for different storage media. Separate, differentiating technical procedures, each with their own set of specialists, will become superfluous to a certain extent.

The combined use and effect of different media such as sound, light, film, video and computer animation in theatre, opera and musical show business is not new in itself. Wagner's vision of a *Gesamtkunstwerk* pointed in exactly the same direction New media theory, based on ideas set forth by Walter Benjamin and Marshall McLuhan, is dealing increasingly with Wagner's and Nietzsche's aesthetic notion of the *Gesamtkunstwerk*. Literary critics and media philosophers such as Friedrich Kittler and Norbert Bolz are trying to explore how man's experience of the world is shaped when it is communicated to the senses via primarily technical channels. Aspects of perception understood as *aisthesis* become especially important in this media aesthetic which is particularly relevant to the discussion of musical culture and society.

The media educationalist Rolf Großmann is probably right in treating this approach's implicit separation of audio and video as obsolete: 'The practice of digital audiovision cannot be pressed into any of the categories arising from the traditional differentiation between the arts. Neither their familiar rank amongst themselves nor their aesthetic premises can be of any use in describing or evaluating the intentions, methods and manifestations of current media praxis, nor in opening up any novel perspectives for the future?

'Even the synaesthetic, intermedial approach to the 'sounds of images' and to the 'liaison of music and visual arts', no matter how courageous and useful it may be, does not go far enough. Although such approaches connect the auditive and the visual spheres, they still fail to integrate the final step: the functional integration of images, text and sound in digital programmes and networks.'[13]

The fear, expounded by Großmann[14] and others, that the power of images reduces music to having no future, is not about to become a reality.[15] However, it is certainly unavoidable that music will tend to fulfil increasingly utilitarian purposes in public life. Großmann even states that the new technology will 'musicalise' itself, since otherwise 'ignoring the auditive dimension ... [would] let a substantial quantity of information potential go neglected'.[16] Music would then be used even more in the same way as is already being done in jingles and signature tunes: for its pure signal effect or as a kind of MUZAK for multimedia applications.

3.3 Third Thesis: Virtual realities become highly important in science, technology, education and art

Simulation methods have already become valuable for modelling in science, and there is no doubt that this development will gain further strength and become even more fundamental in the future. Virtual worlds are built up optically and acoustically and designed artistically. For instance, the author of this article was able to experience a jolting ride on the crests of a three-dimensionally depicted wave representing a musical vibration. In other words, this was audio-visual bodysurfing in a virtual musical world, and this is already possible with a relatively inexpensive Silicon Graphics workstation.

Art draws its essence from the stylisation, abstraction (all the way from prehistoric cave drawings to holographic 3D images), metaphorisation and virtualisation of real world experiences. The world, as we interpret it, is a construction of the mind.

Concepts and sign systems such as our notation system already have digital properties to a certain extent (as Cologne musicologist Jobst P. Fricke has recently shown[17]). Now that we can create and implement virtual worlds that offer almost true-to-life experiences, this perspective gains in conceptual precision and becomes even somewhat threatening.

In future, music may be possibly only an – albeit important – partial aspect of spatial compositions that can be experienced in all their audio-visual aspects. Three-dimensional visual spaces in which you can walk around are not the only project that is being worked on; the acoustical aspect is being taken increasingly into account, since the reality impact of a virtual situation also depends, of course, on the auditive impression that is conveyed at different points in such simulated spaces. Stereophony is only the beginning. Some day, it will be possible to hear a virtual concert chosen at a random concert hall that could even be constructed according to the user's wishes. As a listener, you could then choose your vantage point from anywhere in the hall, for instance by getting closer to an instrument if you were especially interested in following that instrument's individual part.

3.4 Fourth Thesis: Global networking, on the one hand, will lead to the effect that cultures increasingly resemble each other; on the other hand, narrower cultural groups with new divergence criteria are emerging

Regional differences between different cultures will either tend to disappear or will be reformed according to modified needs. For instance, radio stations can now be heard in many different countries due to satellite broadcasting, but maybe this factor has also contributed to the creation of quite a number of new regional stations.

Numerous fan clubs on the internet with members from all over the world already exist, and each day new ones come into existence. But maybe such phenomena as teleworking will also contribute to a revaluation of small town and neighbourhood community life, of local cultural activities where, for instance, live music and improvising together become increasingly more significant.

Processes set in motion by the mass media – democratisation, but also individualisation, maybe also the general levelling of art quality – all of these will continue and possibly become more pronounced. But special quality niche products will be available to interested recipients by means of customised access procedures with an individually calculated cost factor. Instead of viewer ratings, the important measurement of user resonance will become the number of accounts in the networks' individual cultural forums.

3.5 Fifth Thesis: Musical life, musicology, music education and the music industry will undergo radical changes

This is the easiest thesis and at the same time the most difficult of all. Easy to state, because everything is bound to change in some way or another. Difficult, because it is practically impossible to predict, with any reliability, how widespread and how significant each of those changes in the different areas of musical activity will be.

It is well-known and has been shown that the digitalisation of modern instrumental, concert and studio technique has, for the most part, become a standard for the production of commercially relevant popular music and has thus had essential consequences in terms of this music's sound, style and structure.[18] Less noticeable are those gradual changes in artistic activities that are related to working at a monitor in the process of composing, arranging and producing and that have become commonplace to a certain extent. In many areas of production, manual music making is losing importance and the trend is shifting toward advance planning musicianship: programming musical events, the automatisation of various kinds of production process, the algorithmisation of compositional ideas and much more. Even if live music is once more in high demand, it is still undecided whether the current 'unplugged' wave and other similar trends actually represent or could become an authentic counterbalance to automatisation, and whether the dreaded desensualisation of music making will truly be missed in the long run.

The effects of new technologies can already be seen in the methods and the content of musicological research. In Germany, research results and music data banks are becoming available on CD-ROM[19] or as WWW pages on the internet.[20] Interactive sounding questionnaires[21] offer researchers in music sociology a set of new, impressive methods for surveying judgements and opinions. Thought is being given to the possibility of making full use of computer support in managing musicologically relevant material.[22] Computational tools are becoming increasingly important for the analysis of musical structures.

New forms of presentation and teaching will become common in musicology (or also in lectures at AGMM conferences). Some day, most speakers' desks will offer the possibility of downloading scores, music examples etc. from the internet and instantly present them on the overhead display. Maybe the lecturer will tend to be only virtually present by way of a video conference link; this is already routine in television news reports.

Aspects related to music education were already addressed above: it is both doubtful and questionable whether future decision-makers in the area of cultural politics will react to the new challenges presented by global networking, edutainment and learning/teaching with computers. Bear in mind, though, that highly instructive WWW pages can be produced, placed on the Net and updated anywhere around the globe. If, say, German educational institutions are either incapable or unwilling to provide adequate information concerning composition with samples and MIDI patterns, then anybody with a knowledge of English can already obtain the necessary information by consulting the corresponding US pages on the Web.

The manifold changes which have taken place in the area of music as a hobby (to name a few: home recording, MIDI composition at the monitor, the virtual splicing of *.Wav-Samples for background noises or techno patterns) are already an integral part of music at home, a past time occupation which has become computerised.

One area, in which major adjustments have not yet been fully implemented, is that of the so-called music market economy. Whereas production of electronic musical instruments has stagnated in Germany and recording studios are currently reconverting from analogue to digital technology, it is especially the distribution sector which still has decisive changes to face. Music publishing houses will tend to put out specialised periodicals on CD-ROM or, even more up-to-date, as WWW pages. Scientific research, scores and individual orchestra parts will not be printed any more, but will be available for downloading from the Net (for instance, in the form of MIDI data), ready to be printed out on any PC. The same can be predicted for all sorts of teaching materials: not only informative texts, but also music and sound examples can be exchanged between whichever scientists and educationalists are interested and they can then be treated in special discussion groups. Functional music requiring no payments for performing rights has already found its market in multimedia applications. New procedures for copying and distributing music (audio on demand) require fully new strategies for collecting fees and protecting artists' rights.

Finally, considering the fact that industrial society is transforming itself into an information society, the question must be asked what the goals of musical education should be. Even though the vast amount of innovations makes it understandably difficult to provide answers to concerns about the future of musical culture and upbringing, anyone with just a little imaginative capability can foresee that there will be very few areas of musical life that will remain untouched by these changes. This makes the need for answers to these questions even more urgent.

The new media must be included in music education. This is necessary, on the one hand, because they are part of daily life, and, on the other hand, because they determine man's relation to the world in a new way.

Specific research in the area of media technologies could provide the basis required for studying and applying media in music instruction. Media and information technologies transform our whole environment and culture. Therefore, music instruction cannot be viewed as if it were isolated from these influences. What is needed is a scientific investigation project and a music-specific technology assessment of the degree of change caused by computer technology in all forms of musical production, behaviour and views about music and musical production.

As part of the investigation project it would be necessary to assess to what extent computers have become an important factor in school children's imagination of musical sound, and to what extent they are thus affecting their perception. What effects do computer-generated sound structures have, especially on young people's understanding of music? What role do the media play in their concrete experience and reception of music? How does new media technology affect the conditions in which learning takes place?

How can new media be implemented in the course of instruction, especially since music perceived through the media has become an important component of the overall musical experience of everyone in our society?

By using computers, synthesizers, sequencers and samplers, school children can learn to form, modify, analyse and use musical sounds creatively. By applying this in their music-making they learn to understand musical structures and try out methods of arranging (for instance with MIDI), composing (for example with algorithms) and improvising (with automatic accompaniment, etc.). Finally, the application of teaching and learning programmes for general music instruction and possibly also for instrumental instruction should be tried out. The author himself has developed ear training programmes and has discussed their methodic/didactical application.[23]

In this context, it does not make much sense to attempt to revive outmoded behaviourist learning theories of programmed instruction from the 1960s. Rather, with the help of multimedia systems, current didactic knowledge must be incorporated into the corresponding computer programmes. This requires many extra hours of programming and becomes very expensive. Publishing houses are reticent to make the necessary investments due to the fact that it is still unclear to what extent such programmes would/will be applied in the schools. What results is learning programmes written by students of computer scientists untrained in pedagogy, or by game manufacturers. This, unfortunately, makes the decline in educational quality feared by von Hentig a reality.

4. Summary

As a matter of fact, it is no longer a question whether a new cultural technique is emerging and is opening new opportunities for teaching and learning, or whether computers are simply another link in the chain of technological evolution, making us more dependent and creating new problems. For quite some time it has been clear that both of these questions have received a positive answer: yes, a new cultural technique has emerged, and yes, new dependencies have also been created. Our current preoccupation now consists in trying to prudently analyse and comprehend the conditions in lifestyle and culture that are associated with the newest technologies, so that their positive effects are put to the best use and their risks are hopefully recognised and reduced – certainly no easy task for anyone today. This article has tried to show that the area of music, being one of mankind's most important cultural achievements, is by no means indifferent to these problems, but rather is affected by them in an especially forceful way. Musical education will not be able to be defined and pursued without taking technological media omnipresence into account.

The music world, especially, should not fall prone to the mistaken belief that computers, multimedia and the internet are all so technically remote that they have no impact on cultural and artistic perspectives. New media technologies are stepping ever more aggressively outside the technical circle and are about to transform and predetermine our cultural life to an almost unimaginably large extent. By probing new educational strategies we should prepare ourselves for developments that will otherwise overtake us unawares. Once they start to be commercially or politically manipulated, the technological wagon could proceed to roll on without us, obliterating valuable, irreplaceable cultural traditions in its path.

From a technological viewpoint, we are already being confronted (or will be so very soon) with digital media technology (multimedia), with global networking (information highways) and with virtual reality (cyberspace and cyberworlds).

From an anthropological perspective, we will thus not be able to avoid experiencing interaction (with computer based systems of all kinds), communication (through digital, international networks) and simulation (of processes, environments and communities) and we will have to integrate them into our general educational goals. We cannot allow the necessary debate to set in too late – a debate concerning modified conditions of imparting knowledge, of values and of criteria for action and decisions.

Considering these comprehensive transformations, a specifically musical media competence is urgently required. In educational terms this means an appropriate maturity[24] in dealing with music technologies in order to be able to start setting new goals for musical education that can cope with future challenges.

Notes

1 This paper is a revised and shortened version of the essay 'Musikalische Bildung und Neue Medien', *Musikforum*, 1995:12, 40-55.

2 U. Heuser, 'Am Bildschirm allein zu Haus', *DIE ZEIT*, 1995:43, 54.

3 H. v. Hentig, *Die Schule neu denken* (München, 1993).

4 M. McLuhan, *The Gutenberg Galaxy* (Bonn, 1995).

5 G. v. Randow, 'Die meta-orale Weltgemeinde', *DIE ZEIT*, 1995:43, 10.

6 Cf. for example: R. Meyer, 'Veränderte Lebensbedingungen durch neue Medien' in E. Ostleitner (ed.), *Massenmedien, Musikpolitik und Medienerziehung* (Vienna, 1987), 129.

7 B. Brecht, 'Radiotheorie, 1927-1932' in *Bertold Brecht, Gesammelte Werke* (Frankfurt/M., 1967, Vol. 8), 133; cf. J. Stange, *Die Bedeutung der elektroakustischen Medien für die Musik im 20. Jahrhundert* (Pfaffenweiler, 1989), 76.

8 Cf. M. Mathews, 'Foreword', in C. Roads, J. Strawn (eds.), *Foundations of Computer Music* (Cambridge 1985), IX/X; R. Dannenberg, 'Computerbegleitung und Musikverstehen' in B. Enders, S. Hanheide (eds.), *Neue Musiktechnologie. Vorträge und Berichte vom KlangArt-Kongreß '91* (Mainz, 1993), 241-252.

9 'Lenny's Toon', a musical computer game, takes the child into a musical raven's colourful world of sound. The bird offers an array of musical puzzles and small adventures in which the child can emerge victorious by solving musical problems. The programme's highlight is a flashy music studio that does not only permit arranging and mixing a song in different pop music styles, but even lets the child choose the singer, the band members and the synchronous light show, so that by the end of the game an animated cartoon music show is ready to start.

Another game encourages children to paint coloured notes that are then made to sound and ultimately create a tune. Cartoon versions of operas, etc., are already beginning to appear and similar productions are forthcoming.

10 N. Postman, 'Mehr Daten – Mehr Dumme', *Das 21. Jahrhundert. Faszination Zukunft.* GEO extra 1995:1, 69.

11 Cf. J. Mohr, 'Das digitale Klassenzimmer', *Spiegel Spezial*, 1995:3, 115-118.

12 Cf. N. Postman, *Das Technopol* (Frankfurt/M., 1992), for example 150.

13 R. Großmann, 'Zukunftsmusik – Audiomedien als Kunst und Werkzeug', *ZUKÜNFTE*, 1995:6:12, 26.

14 Großmann, op. cit. (note 13), 26.

15 Already in 1971 Klaus Jungk expressed doubts as to whether the technological media could really do any permanent damage to music since, in the long run, they'cannot do without the creative innovation potential offered by music itself'. Cf. K. Jungk, *Musik im technischen Zeitalter* (Berlin, 1971), 128.

16 Großmann, op. cit. (note 13), 28.

17 J. P. Fricke, 'Musik: analog – digital – analog. Digitalisierung und Begrifflichkeit als Norm in einer scheinbar analogen Welt' in B. Enders, N. Knolle (eds.), *KlangArt-Kongress 1995: Vortraege und Berichte vom KlangArt-Kongress 1995 an der Universitaet Osnabrueck*, Osnabrueck 1998 (Musik und neue Technologie, 1).

18 E.g. by J. Stange, *Die Bedeutung der elektroakustischen Medien für die Musik im 20. Jahrhundert* (Pfaffenweiler, 1989).

19 For instance, Christoph Reuter (Cologne) has collaborated with the University of Osnabrueck on an interactive CD-ROM encyclopedia of mechanical musical instruments including sound examples. Furthermore, Chinese musical instruments are being audiovisually catalogued on CD-ROM in collaboration with the Music Institute of China (Dr. Baoqiang Han). Both projects were presented at the 1995 KlangArt Congress in Osnabrueck.

20 For example, the Cologne University Musicology Institute's WWW page permits the viewing and downloading of a *Festschrift* (in honor of Prof. Fricke's retirement). This is probably the first German *Festschrift* on the internet.

21 R. Müller, 'Neue Forschungstechnologie: Der Klingende Fragebogen auf dem Multimedia-Computer' in Enders, Knolle (eds.), op. cit. (note 17).

22 P. Sitter, 'Anmerkungen zum Multimediaeinsatz in der Musikwissenschaft' in Enders, Knolle (eds.), op. cit. (note 17).

23 B. Enders, 'Lehr- und Lernprogramme in der Musik' in H. Schaffrath (ed.), *Computer in der Musik – Über Einsatz in Wissenschaft, Komposition und Pädagogik* (Stuttgart, 1991), 105-130; B. Enders, W. Gruhn, 'Computerprogramme' in R. Weyer (ed.), *Medienhandbuch für Musikpädagogen* (Regensburg, 1989) 277-295; cf. also B. Enders (ed.), *Computerkolleg Musik – Gehörbildung 1-4* (Lernprogramme, Mainz, 2. ed. 1993).

24 Cf. B. Enders, 'Deus ex machina? – Musikelektronik, eine pädagogische Herausforderung', *Musik & Bildung*, 1990:1, 4043.

INDEX

abstraction, 129–30
Acousmonium, 212, *213*
acoustic documentations, 208–9, 221*n*
acoustic ecology, 103
acoustic happenings, 203
acoustic recording horns, 179–80, 182–83, 185*n*
acoustic sound material, 210–11
Acousto-Lectric Company, 193
active oscillator, 45
Adorno, Theodor, 9–10, 20, 21, 201
ADSR (Attack, Decay, Sustain, Release), 71
aeroplanes: and artistic sensibility, 110–11; composers' fascination with, 117; as devastating weapon, 111; escapism motif, 111–12; and heroic figures, 112–14
aesthetic expectations, 20–22, 218
African music, 166, 167*n*
Ailred, Saint, 38
airports, sounds of, 103–4
Amacher, Marianne, 217
American Federation of Musicians, 168, 169, 172, 173*n*
American Piano Company, 88
Ampex, 139, 159–60
Ampico player pianos, 88
amplifiers, 71, 150
Anderson, Laurie, 216, 228
ANS sound synthesizer, 13
Antheil, George: 'Ballet Mécanique,' 85, 121, 128–29, 131; concert events of, 121–22; and Futurism, 126, 128–30, 132; and machine music, 10; on noise, 132; preciseness of, 131; technology affinity of, 117; and time, 133; torpedo radio control system of, 18; and transport technology theme, 106, 112
Armstrong, Louis, 86
Armstrong, Neil, 116
ARP Avatar, 56
ars nova, 39
art: analysis of new genres, 219; combination of visual and acoustic art, 212–13; and life, 99–100, 205–6; radio art, 209, 212–13, 221*n*; and reality, 200; variety of expression, 196
artificially intelligent musical instruments, 228
art music: exploring new genres, 25–26; not as programme music, 118; and sound engineering, 23, 24; space travel as theme of, 117; Western tradition of, 33
assignment systems, 54
audience: participatory roles of, 113, 217, 220; reactions to noise music, 121–22; and social responsibility of artist, 218; sustaining interest of, 212, 221*n*
Audio Engineering Society, 23, 145, 146*n*
Auer, Leopold, 175, 177, 178, 181
aura, 219–20
authentic interpretation, 16, 29*n*

Bach, Johann Sebastian, 32*n*, 74–76, 86, 107, 176, 177
Baillot, Pierre, 174
Banda, Koji, 63, 66*n*
Barber, Samuel, 114
Barthelmes, Barbara, 10, 16
Barzun, Jacques, 9
Baudrillard, Jean, 223
Bayle, François, 212
Beach Boys, 143, 160
Beatles, 75–76, 82*n*, 143, 160, 194, 195, 226
Beauchamp, George, 189, 191, 193, 198*n*
Beaver, Paul, 75
be bop jazz, 159
Beethoven, Ludwig von, 84, 106, 199, 200
Belar, Herbert, 13, 151–52
Bell Laboratories, 18, 19, 21, 214
Benedetti, Dean, 166–67*n*
Benjamin, Walter, 10, 201, 219–20, 232
Benny, Jack, 152
Beriòt, Charles de, 174

Berlioz, Hector, 106, 122
Bernstein, Leonard, 114
Berry, Chuck, 159, 160
Bigsby, Paul, 198*n*
Bijma, Greetje, 101
Bijsterveld, Karin, 10
Birkenholtz, Arcadie, 179
Bischoff, F. W., 209
Blathy, Otto, 188
Blitzstein, Marc, 114–15
blues, 107–8, 158
Boccioni, Umberto, 99
Bode, Harald, 54, 56
Boethius, 40–41
Boie, Robert, 46
Bolz, Norbert, 232
boom boxes, 161, 163, 164, 165
Borden, Dave, 77–78, 81*n*
Bösendorfer Grand Player Piano, *93,* 94
bowed string instruments, 49–50
Bowie, David, 226
Brahms, Johannes, 176
Branca, Glenn, 49
Braverman, Harry, 171
Brech, Martha, 15, 25
Brecht, Bertolt, 20, 113, 226
Brendel, Alfred, 22
bricolage, 102, 105*n*
Brody, David, 171–72
Bruitism, 123, 124, 127, 128
Buchla, Donald, 13, 46, 67, 73–74, 76, 82*n*
Burawoy, Michael, 171
Busoni, Ferruccio, 11, 21, 125, 131, 200–201, 208

Cage, John: compositions of, 219; on conspicuous technology, 204; as futurist, 210–11; gramophone as instrument, 15, 20, 102; howling dogs of, 200; influences on, 122; radio as instrument, 20; unification of genres, 213–14
Cahill, Thaddeus, 11, 17, 125, 131, 207, 208
capitalism, 171
Carlos, Wendy, 75
Carson, Ben R., 151
Caruso, Enrico, 201
Casio keyboard synthesizers, 74
cassette culture, 165–66
cassette tape technology, 161–66
CD-ROMs, 47, 225–26, 234, 235, 238*n*
Celebidache, Sergiu, 22
Chandler, Alfred D., Jr., 170–71
Chapman, Emmett, 49
Charlemagne, 37
Charpentier, Gustave, 98–99, 100
China, pianos from, 63
Chopin, Frédéric, 180
Choralcello, 46
choral music, 36
Christian, Charlie, 189, 191
Christophori, 42
church culture, music in, 36, 37–38
Ciamaga, Gustav, 71
cities: as living environments, 104; and music around 1900, 98–99; role in music and art, 97–98; social and cultural changes in, 101–2; as social organisms, 103; as soundscapes, 102–4
classical music: average recording lengths, 151; best selling records, 75; early attitudes toward recording, 20–22; non-European styles of, 34–35; on player pianos, 84–85; sound engineering, 23, 24; stereo recording of, 140; symbiosis with technology, 33; Western tradition of, 34–36
Clavecin électrique, 46
clavicembalo pian e forte, 42
clavichord, 41–42
Clawson, Dan, 171
Cleveland Recording Company, 136–45, 146*n,* 147*n*
clinkers, 137
closed forms, 219
Collins, Nicolas, 16, 102
Coltrane, John, 107–8
Columbia Broadcasting System, 153
Columbia-Princeton Studio for Electronic Music, 71
Columbia Records, 137, 148–49, 152–53, 154
composers: attitudes toward technology, 20–21, 117; cities in works of, 98–99; Futurist stance of, 10, 126, 128–29; of ma-

chine music, 122; player-piano recordings of, 84; railway themes of, 107–10, 117; and self-playing instruments, 84–85
composition machines, 55
computers: complexity of composition, 211; experimental music, 221*n*; first computer-generated sounds, 14; interaction with, 229–30, 231; musical applications of, 14, 43; player piano as predecessor, 90; rock music software, 55; and sound artists, 102; sound variations offered by, 211
concerts: audience reactions to noise music, 121–22; for player pianos, 87, 94; sound distribution, 212; in virtual reality, 233
conductors, 20–21, 24–25
Confucianism, 61
control, 55–56
controlled feedback, 194–95
controllers, 55–57, 73
Cook, Lawrence, 87
Copland, Aaron, 86, 140
counterpoint, 99, 100
Cowell, Henry, 87, 90
crooning, 20
Cuban rhumba music, 166
Culshaw, John, 24
cultural impact theses: digitalisation, 230–31; global networking, 233; multimedia, 231–32; radical changes to musical life, 233–35; virtual reality, 232–33
cultural protest, 101
Curie, Jacques, 49
Curie, Pierre, 49
cybersex, 227

Dada, 213
Daló, Roberto Paci, 101–2, 105*n*
data gloves, 214, 227–28
David, Ferdinand, 174
Davis, Angela, 108
defense technology, 17–18
de la Motte-Haber, Helga, 12, 213
Delius, Frederick, 98–99
Den, Max, 188
Derrida, Jacques, 172
Deutsche Grammophon, 139
Deutsch, Herb, 70
de Wit, Harry, 101
dial-controlled instruments, 53
diatonic harmony, 40
digital audiovision, 231–32
digitalisation, 230–31, 234
digital media, 225–26, 229–30
digital technology, 20, 24
dirty guitar music, 196
disc jockeys, 16
Disklavier, 47
distortion, 152
dobro, 49
Doesburg, Theo van, 134*n*
Dolby noise reduction systems, 162
Domselaer, Jacob van, 128
Donné, Etant, 101
double-tracking vocals, 160
Dowd, Tom, 24, 144
Driscoll, John, 217, 219
Duchamp, Marcel, 106, 111, 205
Dudley, Homer, 18
Dudon, Jacques, 51
Dunstable, John, 39
Dylan, Bob, 160
Dynamophone, 11, 125, 131, 201, 207

Eberhardt, Siegfried, 175, 182
E-bow, 49
echoing, 160
Echo Piano, 55
Eckert, Franz, 60
Edison, Thomas, 159
editing, 139
education, 59, 64–65, 235–36
edutainment, 229, 234, 237–38*n*
Edwards, Richard, 171
Eimert, Herbert, 200, 210
Eisenmann, Richard, 46
electric guitars: controllers for, 56; development of, 49; early electric models, 189; electromagnetic pickups, 193; empowerment of marginal groups, 195–96; feedback, 12, 186–87, 191, 193–95; history, 188–91; loss of sound control, 188; musician freedom from big bands, 17; new role of, 15; noise as

electric guitars (*continued*)
music, 188; popularity of, 12, 195; solid-body models, 187, 188, 191, 193–94; as symbolic vehicle, 196; synthesizers compared to, 43, 198*n*; technological innovation, 186–87
electronic instruments: about, 52–54; Buchla on, 73; classification system, 43–44; defined, 44–45; development of, 10–14; electro-acoustic instruments, 44, 45, 47–50, 210–11, 212; electromechanical instruments, 45, 50–52; electrophones, 44–47; principle types of, 55
electronic music: and creativity, 15–17; Eimert's definition of, 200; loathed by composers, 16–17; *musique concrète* merged with, 202; terminology of, 44
electronic music studio, 56
electronic percussion, 56
Electronic Sackbut, 56
electronic timpani, 49
electrophones, 44–47
electrostatic principle, 48–49
electrostatic tone-wheels, 51–52
'elektronische Musik,' 210
Elektrophonisches Klavier, 46
Elgar, Edward, *19,* 98–99
Ellington, Duke, 107, 109, 115–16
Elman, Mischa, 180, 182
Emerson, Keith, 76, 82*n*
Emerson, Lake, and Palmer, 76
Enders, Bernd, 14
enharmonicism, 125–26
Eno, Brian, 77, 103
ensemble singing, 39
envelope shapers, 69
environmental sound, 103
equalizers, 139, 141
eschiquier anglais, 41
Esquire guitar (Fender), 194
Estrada, Julio, 87
experimental music, 221*n*

Fab 5 Freddy, 163
Fachiri, Adila, 178
faders, linear, 139
Fairlight, 55
Fast Forward, 102
Federal Communications Commission, 151
feedback: acceptance of, 194–95; controlled feedback, 194–95; in electric guitars, 12, 186–87, 191, 193–95; as noise, 191, 193–95
Feldman, Morton, 16
Fender guitars, 194
Fender, Leo, 194, 198*n*
Fenton, John William, 59
fidelity: advances in, 22; early limitations, 20; high fidelity, 149, 151–53, 155–56*n*; of long-playing (LP) records, 148
fingerboards, 53
First World War, 111, 112
Fisher, Linda, 79
Flesch, Carl, 175, 176, 177, 179, 181, 182
flight themes, 113–15
Fluxus movement, 213
Fontana, Bill, 103, 205
45-rpm records: contribution to rock music, 154; engineering, 151; introduction of, 148, 152; markets for, 153; RCA's intentions for, 153–54; sound quality of, 153
Foucault, Michel, 172
Francescatti, Zino, 177–78
frequency, improvements in range, 19
Fricke, Jobst P., 233
Frith, Fred, 49
Fromm Music Foundation, 39–40
Frying Pan guitar, 189, 193
Furtwängler, Wilhelm, 21
Futurism: and abstraction, 129; aeroplane themes, 111; breaking laws of nature, 126; and cities, 99–100, 102; and composers, 10, 126, 128–29, 200, 210–11; concerts of, 106; engagement with life, 123; and machines, 122, 132; music freed from repetition, 127–28; noise music of, 199–200; technology embraced by, 123–25, 204; themes of, 111

Gabriel, Peter, 166, 226, 227
Gaisberg, Fred, 24
Galilei, Vicenzo, 41
Galpin, Francis, 44
Gates, Bill, 226
Gatty, Nicholas, 126

Gaulard, Lucien, 188
Geertz, Clifford, 196
genres: acousmatic music, 212; acoustic documentations, 208–9, 221*n*; closed forms, 219; electro-acoustic music, 210–11, 212; 'elektronische Musik,' 210; emergence of new genres, 207; 'Hörfolgen,' 208–9; radio art, 209, 212–13, 221*n*; sound installations, 216–18, 219; sound performance, 213–14; sound sculptures, 215–16, 219; traditional intermingled with new, 219; unification of, 213–14
German technology, 139, 159
Ghazala, Qubais Reed, 46
Gibbs, John, 188
Gibson guitars, 189, 191, *192*
Gibson, William, 101
Gillespie, Dizzy, 159
global information network, 225, 226–27
global networking, 233
Gnazzo, Anthony, 71
Goodman, Benny, 189
Gore, Al, 225
Gould, Glenn, 21, 23
gramophones, 15, 130–31, 201. *See also* phonographs
Grand Funk Railroad, 142
granular synthesis, 231
GRM *(Groupe de Recherches Musicales),* 109–10
Großmann, Rolf, 232
Gronostay, Walter, 209
Grubb, Suvi Raj, 24
Guido d'Arezzo, 39
Guitaret, 50
guitars. *See* electric guitars; Hawaiian guitars; Spanish guitars; steel guitars
Guitar Slim, 193, 194

Haass, Hans, 85
Haley, Bill, 141, 142, 160
Hall, Marie, 177, 182
Hamamatsu, Japan, 61, 65
Haman, Gloria Busse, 136
Haman, Ken, 136, 137–45, *138, 140,* 146*n*, 147*n*
Haman, Paul, 147*n*
Hammond, John, 189
Hammond, Laurens, 11–12
Hammond organ, 11–12, 51
Hanert Electrical Orchestra, 13
Hanert, John, 13
Hansen, John, 142–43
harmonica, 43
harpsichord, 42
Harris, Frank, *68,* 72
Harrison, George, 75–76, 82*n*
Hart, Gary, 191
Hauck, Werner, 178
Hawaiian guitars, 189, *190,* 191, 193
Heifetz, Jascha, 177, 180, 182
Hendrix, Jimi, 15, 194
Henry, Pierre, 201, 202
Heuser, Uwe Jean, 223
high fidelity, 149, 151–53, 155–56*n*
hillbilly music, 195
Hiller, Lejaren, 14
Hill, Gary, 216, 219
Hindemith, Paul, 85, 113, 114, 131, 207
Hindley, Geoffrey, 9
hip-hop, 163–64
hire purchase systems, 65
Holly, Buddy, 160
holographic detectors, 53
home recording studios, 164–65
Honegger, Arthur, 106, 107, 113–14, 122
Hooker, John Lee, 226
'Hörfolgen,' 208–9
Hornbostel, Erich von, 43, 55
Horning, Susan Schmidt, 23
Horowitz, Vladimir, 84
Huberman, Bronislaw, 180
Hughes, Thomas, 187–88
Human Beinz, 142, 143
humanistic ideal, 40–41
hurdygurdy, 38–39
hydraulis organ, 36–38

impressionism, 99
India, 34–35, 165
information highways, 226–27
information society, 224–25

ı detectors, 53
ıctive sound installations, 217–18
ıedia art performances, 213–14
ınet: global networking, 233; as informa-
ıon society, 224; music forums on, 227;
surfing, 229
ıterval, 34, 35
ıntonarumori, 100, *121*, 123–24, *124*, 199–200, 208
Isaacson, Leonard, 14
Italian Futurism. *See* Futurism
Ives, Charles, 98–99
Izawa, Shuji, 59-60

James Gang, 142, 143
Janney, Christopher, 46
Japan: cassette tape manufacturing, 161, 162; education system modernization, 59; mass production, 64, 65; organ production, 60–62; piano production, 61–62, 65*t*; piano tuning and moisture, 62–63; precision of pianos, 63–64; traditional musical styles, 35; Western instruments introduced, 59–60; Western style classical music tradition, 33; work ethic in, 61
jazz: archival records of, 159; be bop, 159; groups as microcosmic utopias, 116; mass audience of, 195; railway themes, 107–8; recording, 21, 158; removed from harmony, 127; sound engineering, 24
Jermée, Paul, 209
Joachim, Joseph, 174, 176, 177, 178
Johnson, Eldridge, 149
Johnson, Howard, 113
Johnson, Lonnie, 189
John VIII, Pope, 37
Jolivet, André, 207
Joplin, Scott, 110
Jordan, Stanley, 49
Joyce, Thomas F., 151, 155*n*
Joy, Lenny, 139
Jungk, Klaus, 238*n*

Kagel, Mauricio, 203
Kandinskly, Wassily, 99
Karajan, Herbert von, 21
Kästner, Erich, 208–9
Katz, Mark, 21
Kaufman, Louis, 179
Kawai, Koichi, 60
Kawai Piano Company, 61, 65
Kawakami, Genichi, 62, 63
Kazdin, Andrew, 24
Kealy, Edward R., 144
Keller, Hans, 178
keyboard instruments, 41–42, 46
keyboards: alternatives to, 82*n*; for MIDI, 56; of player pianos, 89; as remote control, 56; split keyboards, 47, 54; for synthesizers, 73–74
keyed percussion instruments, 50
Khaled, Cheb, 166
Kimbara, Meizin, 61
King, Tom, 141
Kittler, Friedrich, 232
Klemperer, Otto, 22, 24
Klüver, Billy, 214
Knapp, Orville, 191
Knight, Terry, 142
Kobayashi, Tatsuya, 14
Kock, Winston, 18
Koenig, Gottfried Michael, 14
Korea, 33
Koussevitsky, Serge, 114
Kraft, James, 17
Kraftwerk, 107
Kreisler, Fritz, 176, 179, 182
Krenek, Ernst, 131
Kubelik, Jan, 177
Kubisch, Christina, 104

labor legislation, 169, 170
labor unions, 168, 169, 172, 173*n*
Lady's Glove, 214
Lamarr, Hedy, 18
Lang, Eddie, 189
language, shaping social reality, 172
lap steel, 189, *190*, 191, 193
Lea Act, 169
Le Caine, Hugh, 56
Lee, Spike, 160
Léger, Fernand, 129
Legge, Walter, 24, *25*
Lemberg, Götz, 217

Lemon Pipers, 142
Lennon, John, 82*n*
Leoninus, 39
Lévi-Strauss, Claude, 16, 97, 102
Lewis, Meade 'Lux,' 109
liberation, 17, 107–8, 116
life, 99–100, 123, 205–6
Ligeti, György, 94
limiters, 139
Lindbergh, Charles, 113, 226
Lindy Hop, 113
linear faders, 139
Linn Drum, 56
Lisitzky, El, 97
Liszt, Franz, 84, 181, 182
live recording, 22
Loar, Lloyd, 49, 193
Lockert, Jimmy, 139
longplaying (LP) records, 148, 150, 152–53
Lopatnikoff, Nicolai, 85
Louis the Pious, 37

machine music, 9–10, 130–32
Mack, Wayne, 137
Mager, Jörg, 54, 201
magnetic recording, 159
Magoun, Alexander, 10
Mälzel, Johann Nepomuk, 84
Manuel, Peter, 165
Marclay, Christian, 16
Marinetti, Tommaso, 111, 123, 124
Marley, Bob, 227
Martinu, Bohuslav, 114
Marx, Karl, 171
Mason, Luther W., 60
mass production, 64, 65, 74, 149, 163
Mathews, Max, 14, 46, 50
Matsumoto, Shinkichi, 60
Matsushita, 162
Matzunaga, Sadajiro, 60
McLuhan, Marshall, 225, 232
McSwain, Rebecca, 12
mechanical instruments, 46–47, 50, 84–85
media, revolution in, 101
Meidner, Ludwig, 97
Meier-Dallach, Hans-Peter, 103
Meier, Hanna, 103
memory storage devices, 47
Memphis Minnie, 193
Mervar, Anton, 137
Messiaen, Olivier, 207
Metheny, Pat, 195
metropolis myth, 97, 104
microphones: attachable, 50; built-in, 48, 50; contact type, 50; early technologies, 18–20; equalizers for, 141; external, 48; influence on violin vibrato, 179–80, 182–83
microtones, 125–26, 129, 134*n*
MIDI: composing, 226, 234; critics of, 15; with data gloves, 214, 228; as digital control system, 56; in electric instruments, 50; establishment of, 14; file swapping, 227; keyboards for, 56; quantification of music, 230; reduction of musical information, 231; sequencing, 226; traditional instrument controllers, 55
Miessner, Benjamin F., 11, 17–18, 195
military technology, 17–18
Millard, Andre, 24
Miller, Glenn, 109
Miller, Jack, 191
Minard, Robin, 218, 219
miniaturisation, 150
minimal music, 25
Mini-Moog, 74, 77, 80
Miranto, James, 113
Mitchell, Joni, 166
Mixturtrautonium, 56
modernism, 102
modern lyre, 41
Monahan, Gordin, 102
Mondriaan, Piet, 10, 122, 126–28, 129–33, 134*n*
monophonic instruments, 47, 52–53
montage-like world views, 201
Montgomery, David, 171–72
Moog, Robert A., *68*; background, 70; and Beatles, 82*n*; Bode's influence on, 56; Buchla compared to, 73–74, 82*n*; and controller technology, 73; current pursuits, 80; and Deutsch, 70–71; education, 81*n*; and Mother Mallard, 79; musical ability of, 81*n*; popularity of synthesizer, 74–77; production of early synthesizers, 71–72; promotional material by, 72; on responding to demand,

Moog, Robert A. (*continued*) 198*n*; on 'Switched On Bach,' 75; synthesizer production begun, 13; on transistors, 69; Trumansburg workshop of, 67
Morton, Jelly Roll, 86
Mosolov, Alexander, 122
Moss, David, 101
Mother Mallard, 72, 79
mouth organ, 43
movies: effect of sound movies on musicians, 168–69; player-piano accompaniment of silent films, 85; synthesizers in, 76, 82*n*, 83*n*
Mozart, Wolfgang Amadeus, 85, 122, 199, 215
multimedia, 225–26, 231–32
multi-track capabilities, 20, 142, 143, *144*, 145
Mumford, Lewis, 172
Murzin, Evgenij, 13
music: in church culture, 36, 37–38; and cities around 1900, 98–99; city roles in, 97–98; copyright protection of, 160; cultural impact of digital technology, 233–35; data banks, 234, 238*n*; digitalisation of, 234; dirty guitar music, 196; education in new media, 235–36; emergence of new genres, 207; experimental music, 221*n*; extension of material, 199–201; forums on internet, 227; in legend, 34; man-machine debate in, 130–32; neo-plasticism, 127–28; new technologies, 102; noise in, 100, 110, 121–22, 123–25, 129, 132; notation, invention of, 39; oligopolistic industries, 170; physicality of, 34; publishing on web, 235; as socio-cultural product, 68; technique development, 9–10; time as canvas of, 128; time space components, 112; unplugged, 25; Yamaha's role in education, 64–65
Musical Instrument Digital Interface. *See* MIDI
music engineers, 17–18
musicians: African Americans, 195; cassette as medium of choice, 162–63; experiments with nonmusical sounds, 194–95; expression loss during recording, 181–82, 183; golden age for, 168; innovation accepted by, 170; magnetic recording's influence on, 158; as poor paying customers, 143; as recording engineers, 161; relationship with recording engineers, 141; as synthesizer users, 72, 74–77, 78–79, 82*n*; tape recording by, 159–60; and technological changes, 168–69
music market economy, 234–35
Music Research Institute, Japan, 59–60
musique concrète: acoustical documentations, 221*n*; concerts, 212; electronic music merged with, 202; electronic pioneering, 15–16; everyday sounds of, 109–10; origin of, 12, 209–10; sampling, 200
musique d'ameublement, 208, 216

Nakamichi, 162
Nancarrow, Conlon, 11, *85*, 85–94, *88*
Nathan, Sydney, 152
Negroponte, Nicholas, 224
neo-classicism, 129
Neo-Dada movement, 213
neo-plasticism, 127–28
Neuhaus, Max, 217
Neumann, George, 139
new media, 225, 235–36
New Music, 39–40
New Music Edition, 86, 87
New Objectivity, 10
new vibrato, 174–78
new wave music, 163, 164
Nick, Edmund, 208–9
Nietzsche, Friedrich, 232
Nishikawa, Torakichi, 60
Nishikawa, Yasuzo, 60
Nobel, David, 171
noise: Antheil on, 132; feedback as, 191, 193–95; in Medieval mystery plays, 199; in music, 100, 110, 121–22, 123–25; Russo on, 199–200
noise instruments, 127. *See also* intonarumori
noise machines, 204
noise music, 199–200
noise reduction systems, 162
nonlinear distortion, 152
Nono, Luigi, 204
Norelco, 161

objectivity, 131
Odington, Walter, 39
Ohlischlaeger, Genö, 209
Ojima, Naokichi, 60

oligopoly, 170–71
Olson, Harry, 13, 151
on-demand media, 223, 224
Ondes Martenot, 53, 200, 207
orchestras: presumed demise of, 17, 130–31; samplers used in, 44; synthesizers used in, 17, 44, 211–12
orchestrion, 84
organistrum, 38
organs: advent of, 36–38; arrival in West, 37; electric reed organs, 50; electronic, 11–12, 18; invention of, 36; Japanese production of, 60–62; keyboard evolution, 41–42; Yamaha and, 33
Ornstein, Leo, 111, 117
Orpheus, 34, 35, 41
oscillators: beat-frequency, 52; development of, 43; electromechanical, 51; in electronic instruments, 52; electrostatic, 48; master oscillator, 54; in monophonic instruments, 53; passive electroacoustic, 47–48; in polyphonic instruments, 54; shared, 47; types of, 45; voltage controlled, 69, 70
Ostertag, Bob, 231
Oud, J.J.P., 134*n*
Ozawa, Seiji, 64

Paley, William S., 152
Panasonic, 161
paper tape recording, 159
Parker, Charlie, 159
Parkinson, Donald B., 18
Parmegiani, Bernard, 109–10
partially polyphonic instruments, 53–54
Partiturophone, 54
passive oscillator, 45
patriotism, 113, 114
Paul, Les, 49, 194
Paxton, Steve, 214
Perelman, Lewis, 229
Pere Ubu, 143, 145, 147*n*
Perkins, William, 151
Perotinus, 39
Perry, Matthew Calbraith, 59
Petrillo, James C., 169
Philicorda, 52
Philips Company, 161
Philips, Sam, 139, 146*n*
phonographs. *See also* gramophones: automatic changers, 150; invention of, 201; as musical instruments, 130–31; radio combined with, 149–50; reliability of, 154; sound quality of, 154
photoelectric instruments, 48–49
photoelectric principle, 51
piano attachments, 55
Pianocorder, 47
pianos. *See also* player pianos: Chinese, 63; Echo Piano, 55; electric, 49; electronic, 11; invention of, 42; Japanese production of, 61–62; piano player *vs.* player piano, 89–90; precision of Japanese, 63–64; production of Japan *vs.* United States, 65*t*; sustaining, 49; tuning considerations in Japan, 62–63; Yamaha and, 33
Piatti, Ugo, *121*
piezoelectric principle, 49, 51
Pijper, Willem, 132
Pinch, Trevor, 13–14
piracy, 165
Piston, Walter, 86
Plato, 34
player pianos: about, 11; Bösendorfer Grand Player Piano, *93*, 94; classical music on, 84–85; composition process, 90, 92; as computer predecessor, 90; hammers, 94; invention of, 84; keyboards, 89; limitations of, 88–89; metre, 90; modern forms of, 46–47; Nancarrow's compositions for, 90, *91*, 92–94; Nancarrow's path to, 86–87; *vs.* piano players, 89–90; possibilities of, 88–90; principle of, *89*; rhythm, 90; speed of, 89–90
polka, 136–37
polyphonic instruments, 47, 54
polyphony, 41
popular music, 23–24, 159–60
Postman, Neil, 229, 231
postmodernism, 101
post-structuralism, 172
Power, Leonel, 39
Praetorius, Michael, 199
Pratella, Francesco Balilla, 111, 125
precision, 63–64, 130, 131
Presley, Elvis, 152, 160

Presto autodisc system, 158
Preston, Don, 76
Prieberg, Fred, 212
Prince, 226
problem-solving, 187–88
Prokofiev, Sergej, 122
punk, 101, 164

racial prejudice, 116
radio: audience participation, 113; early technologies, 20; effect on musician market, 169; FM licensing, 151; in high fidelity, 151; 'Hörfolgen,' 208–9; as instrument, 20; military application, 18; oligopolistic industries, 170; phonograph combined with, 149–50; three-channel broadcasting, 140
radio art, 209, 212–13, 221*n*
ragas, 35
railways: composers' fascination with, 107–10, 117; rhythmic noise montages, 202; stations sounds, 103, 109; train accidents in music, 110
Randle, Bill, 141
Ranger, Richard H., 11, 17–18
Rangertone Organ, 18
rap, 163–65
Ravizza, Guiseppe, 42
Ray, Man, 106
RCA Electronic Music synthesizer, 13, 51, 55
RCA Victor: competitors of, 148–49, 150, 153; high fidelity, 149, 151–53; intentions for 45-rpm records, 153–54; introduction of 45-rpm records, 148, 152; leadership position of, 155*n*; longplaying (LP) records, 153; sales boom after Depression, 149–50; standardized recorded music systems, 150; and television, 150, 152
reactive sound installations, 217–18
real time processing, 214
recording. *See also* tape recording: on acetate discs, 158; advances in, 22; as art form, 144; cassette's effect on hegemony, 165; as catalyst for change, 183; Cleveland Recording Company, 136–45; console innovations, 139–40; double-cassette decks, 164; early technologies, 18–21; effect on musician market, 169; as evolutionary process, 141–44; exigencies of, 179–83; 45-rpm records, 148, 151–54; hegemony of recording companies, 158, 160, 161; high fidelity, 149, 151–53, 155–56*n*; home recording studios, 164–65; improvements in technology, 185*n*; influence on music making, 21–22, 158; long-playing (LP) records, 148, 150, 152–53; loss of musician's expression, 181–82, 183; magnetic recording, 159; oligopolistic industries, 170; playback systems, 149; rock music evolution, 141–42; session durations, 137; 78-rpm records, 148, 150, 153; standardized recorded music systems, 150; state-of-the-art equipment, 145; stereo classical music, 140; technology's impact on, 136; violin vibrato transitions, 175–78
recording engineers, 137–38, 141, 144, 145, 161
Recording Industry Association of America, 147*n*
reed instruments, 50
Reichel, Hans, 49
Reich, Steve, 110
remote control, 56
repetition, freedom from, 127–28
re-recordable CDs, 47
Reuter, Christopher, 238*n*
reverse engineering, 59, 63–64, 65
reverse salient, 187–88, 191, 193–95
rhythm, 39, 90, 127–28, 202
ribbon controller, 73
Ricci, Matteo, 34
Richter, Svatoslav, 64
Rickenbacker company, 49, 193
Rigby, Helen, 86
ring modulator, 77
Ritter, Don, 217, 219
rockabilly, 158
rock music: demands on technology, 141; devices supported by, 43; evolution of recording techniques, 141–42; 45-rpm record's contribution to, 154; irreproducible live sounds, 143–44; software for, 55; tape recording's advantages for, 158
Rose, John, 101
Rosen, Theodore, 39
Rossini, Gioacchino, 110
Rubin, Joel, 101–2, 105*n*

Russolo, Antonio, 127
Russolo, Luigi, *121*; about, 125; and Bruitism, 124; concert events of, 121–22; enharmonicism of, 125–26, 128; and Futurism, 99, 126, 127, 132; on futurist noise music, 199–200; noise instruments of, 100, 123–24, 132, 208; qualities sought by, 129–30; and reality, 133
Rutters, Herman, 132
Ruttmann, Walther, 201, 208

Saariaho, Kaija, 211–12
Sachs, Curt, 43, 44, 55
Saint-Exupéry, Antoine de, 113–14
Saita, Mitsunori, 60
Saito, Hideo, 64
Sala, Oskar, 56
sampling: appearance of, 14; and cassette recorders, 165; concept developments, 201; early forms of, 200, 201; forerunner of, 52; of images, 202; lost time coordinates, 205; memory storage increases, 55; and montage, 201; and perceptual changes, 204–5; relationship between art and reality, 200; spread of, 43–44, 51
Sanders, Paul, 134*n*
Saraga Generator, 53
Sarasate, Pablo de, 176, 177–78, 180
Sarnoff, David, 150, 151, 152, 155*n*, 156*n*
Satie, Eric, 208, 216, 220*n*
Saunders, Archibald, 174
saxophone, 15
scales, systems of, 35
Schaeffer, Pierre, 109, 122, 202, 205, 209–10, 212
Schafer, R. Murray, 10, 103
Schaffer, Myron, 71
Schikanaeder, Emanuel, 215
Schillinger, Joseph, 207
Schönberg, Arnold, 20, 122, 208
Schumann, Robert, 181, 182
Schütz, Heinz, 32*n*
Scott, Jim, 77
Scriabin, Alexander Nikolayevich, 13
Sear, Walter, 70, 75
Second World War, 114–15, 117
Seidl, Toscha, 177
Seley, Jason, 70, 81*n*
self-playing instruments, 84–85. *See also* player pianos
Serafin, Tullio, *25*
Serrano, Jacques, 46
Sessions, Roger, 86
78-rpm records, 148, 150, 153
Severini, Gino, 123
Severn, Edmund, 181
shared oscillators, 47
Sherman, Al, 113
Sholpo, Evgeny, 13, 55
Siegmeister, Elie, 114
signal processors, 56
Simon, Paul, 166
Slingerland Company, 193
Slonimsky, Nicolas, 86
Smith, Bessie, 86, 107
social change, 224–25
Social Construction of Technology, 67–68
social flux, 196
social reconstruction, 188
Society of Professional Audio Recording Services, 145
Sonami, Laetetia, 214, 220
Sony, 161, 162
sound engineers, 23, 24–25
sound generation, 46
sound installations, 216–18, 219
sound objects, 204–5
sound performance, 213–14
sound poems, 213
sound production, 46
soundscape, 10, 102–4
sound sculptures, 215–16, 219
Southwort, Michael, 10, 103
space-controlled instruments, 53
space travel themes, 115–17
Spanish guitars, 189, 191, 193
spherophone, 201
Spohr, Louis, 174
steel guitars, 49
STEIM, 214
Steinway pianos, 63
Steinway, Theodore, 63
Stelarc, 101
stereo: broadcasting, 139; classical music recording, 140; recording, 20, 139

Stockhausen, Karlheinz: on classical composers, 32*n*; compositional tools of, 16; on consistency, 29*n*; and electronic revolution, 25; 'Helicopter Quartett,' 115; sound materials of, 12–13; 'Studie II,' 210; 'Telemusik,' 202–3, 204
Stokowski, Leopold, 21, 155–56*n*
Stowell, Robin, 178
Stratocaster, 194
Strauss, Botho, 223
Stravinsky, Igor: authentic interpretation of, 16; Futurists influence on, 200; Nancarrow influenced by, 86; and neo-classicism, 129; reaction to concerts of, 106, 122; recordings of, 140; 'Sacre,' 226; Western concept of music, 35
Strayhorn, Billy, 109
Stuckenschmidt, Hans, 130–31, 132, 133
studio synthesizers, 67
sub-mixing, 139
Subotnick, Mort, 76
Sugar Hill Gang, 163
Suma Recording, 143, 147*n*
Sun Ra, 77, 116, *116*
surfing, 229
surrealism, 202
Suzuki, Akio, 104
swing, 107, 109
'Switched On Bach,' 74–77
synclavier, 14, 55
synthesizers: analogue models, 53, 54; by Buchla, 73–74, 76, 82*n*; classification of, 54; controllers for, 56; culture of, 78–79; development of, 13–14, 56; electric guitars compared to, 43, 198*n*; first use of term, 55; horizontal market integration, 15; invention of, 67; keyboards for, 73–74; monophonic models, 52; by Moog, 13, 67, 69–79; in movies, 76, 82*n*, 83*n*; in orchestras, 17, 44, 211–12; partially polyphonic models, 47; sociology of, 67–68; spread of, 43–44; unpredictability of, 77–78; Yamaha production of, 62
Szigeti, Joseph, 177

Taft-Hartley Act, 169
tape recording: advantages of, 158; cassette technology, 161–66; initial use of, 139; by musicians, 159–60; and popular music, 23–24, 159–60; technology advances in, 160; widespread use of, 160
technology: aesthetic changes of reality, 204; capital's use of, 171; composers not critical of, 117; contextualist view of, 172; creative adaptation to new needs, 145; cultural impact theses, 230–35; destructive potential of, 111–12, 115; dissolution of space-time coordinates, 205; heroism of, 109, 112–13; impact on sound recording, 136; move toward new, 102; natural environment altered by, 204; perceptual changes rendered by, 204; relationship with culture, 117–18; rock's demands on, 141; symbiosis with classical music, 33; tape recording advances, 160; workers impacted by, 168–73
technology transfer, 65
Telemusik, 202–3
teleshopping, 227
television, 150, 152, 169
teleworking, 233
Telharmonium, 11, 17, 51
temporal dissonance, 93
Termen, Lev, 11, 17–18, 207
Theremin, 11, 46, 53, 70, 80, 200, 207
33 1/3-rpm records, 148, 150, 152–53
Thompson, Kay, 140
timbre assignment, 54
time-music, 129
'Tipoo's Tiger,' 215, *215*
Toch, Ernst, 85, 131
Tokyo National University of Fine Arts and Music, 59–60
tonality: breaking off tones, 127, 128; experimentation with guitars, 186; individualization of tone, 182; microtones, 125–26, 129, 134*n*; modern systems of, 130; violin vibrato hiding imperfect intonation, 180–81
tone generators, 114
tone painting, 98–99
tone-wheel instruments, 50–52
Torme, Mel, 140
tornavoz, 186
Toronto Electronic Music Studio, 71
Torres, Antonio, 186
Toscanini, Arturo, 21

Tracy Twins, 140
trains. *See* railways
transducers, 48–49, 50
transistors, 69, 162
transport technologies, 106, 117–18. *See also* specific technologies
Trautonium, 200, 207
Trautwein, Friedrich, 207
Trocco, Frank, 13–14
tuning, of pianos, 62–63
Turkish Janissary bands, 199
typewriter, invention of, 42

ultrasonic detectors, 53
unplugged music, 25
urban aboriginals, 100–102
urbanization, 97–98
Ussachevsky, Vladimir, 12, 71

Valéry, Paul, 203
Van Dijk, M., 128
van Gelder, Rudy, 24
Varèse, Edgar, 10, 11, 14, 122, 126, 200, 210-11
variophone, 13, 55
vaudeville, 168
vibrato, 21–22. *See also* violin vibrato
Victor Talking Machine Company, 148, 149. *See also* RCA Victor
video clips, 231
video conferencing, 227
Vieuxtemps, Henri, 176
Villa-Lobos, Heitor, 109
Vinylite records, 149, 151
violins: bow, 174, 183*n*; chin rest, 178; electric, 50; finger placement, 181; Japanese production of, 60; metal *vs.* gut strings, 178–79; playing aesthetics, 21–22
violin vibrato: and acoustic recording horns, 179–80, 182–83; as camouflage for bad technique, 181; chin rest influence, 178; and finger placement, 181; hiding imperfect intonation, 180–81; individualization of tone, 182; metal string influence, 178–79; and microphones, 179–80, 182–83; in response to recording exigencies, 179–83; transition to new vibrato, 174–78
Virilio, Paul, 223
virtual reality, 223–25, 227–28, 229–30, 232–33
visual elements, sound material, 219
visually-supported audio art, 212
Vitry, Philippe de, 39
Vivaldi, Antonio, 199
Vivi-Tone guitar, 193
vocoder, 18
Vogel, Vladimir, 117
voices, 92–93
voltage control, 56, 69, 71, 82*n*
voltage controlled filters, 77
von Hentig, Hartmut, 223–24
von Huene, Stephan, 215–16

Wagner, Richard, 39, 200, 232
Waisvisz, Michel, 46, 57, 227
Wakeman, Rick, 76
Walker, T-Bone, 189, 191
Walkman, 161
Walsh, Joe, 142, 143
war. *See* First World War; Second World War
Warbo Formant-Orgel, 54
Warren, Harry, 109
Watson, Ben, 83*n*
waveforms, 51–52
waveshapes, 51
WDOK, 137–38, 139
Weber, Max, 9
Weibel, Peter, 217
Weil, Irving, 132
Weill, Kurt, 113, 114
Weiss, Jon, *68*; about, 72; on concerts, 78–79; on keyboards, 73; on Sun Ra, 77; on 'Switched On Bach,' 75; on synthesizers, 72, 76, 77–78, 79
Welte, Edwin, 52
Welte-Mignon pianola, 131
Wexler, Jerry, 144
White, Don, 142
White, Lynn, Jr., 38–39
Whitesitt, L., 129
Wild Cherry, 143, 147*n*
Williams, Andy, 140
wind controllers, 56–57
Winkler, Justin, 102
wire recording, 159
Wolff, Dick, 214

Wolf, Fred, 136–37, 139, 140, 141, 142, 146*n*
Woodstock, 194, 195
work environments, 171
world music, 203
World Wide Web: edutainment, 234; music data banks on, 234; music forums, 227; periodicals on, 235; surfing, 229
Wright brothers, 115

Xavier, Francisco de, 59
Yamaha Corporation: DX7 synthesizer, 74; Electone organ, *62*; first organ made by, *61*; history of, 59–64; keyboard synthesizers of, 80; music class in Germany, *64*; role in music education, 64–65; subsidiaries of, 62; synthesizer production, 62
Yamaha, Torakusu, 33, 60, *61*
Yankovic, Frankie, 136–37, 146
Ysaÿe, Eugène, 176

Zairan *soukous* music, 166
Zipernowski, Charles, 188
Zorn, John, 102

AUTHORS' VITAE

BARBARA BARTHELMES is a musicologist who specializes in contemporary music, microtone music, sound art, synaesthesia and videoclips. She lectured at the Hochschule der Kuenste in Berlin and at the University of Bielefeld, was co-editor of the journal *Musica* and since 1991 has been a member of the editorial board of the music journal *Positionen*. From 1995 to 1998 she was a member of the executive committee of the Institut fuer Neue Musik und Musikerziehung Darmstadt. Her dissertation on the musical and theoretical œuvre of Ivan Wyschnegradsky, *Raum und Klang: das musikalische und theoretische Schaffen Ivan Wyschnegradskys,* was published by Wolke, Hofheim, in 1995.

KARIN BIJSTERVELD is a historian and associate professor at the Technology and Society Studies Department, Faculty of Arts and Culture, University of Maastricht. Her dissertation is titled: *No matter of age. Welfare state, science and debates about the elderly in the Netherlands, 1945-1982* (Geen kwestie van leeftijd. Verzorgingsstaat, wetenschap en discussies rond ouderen in Nederland, 1945-1982, Amsterdam: Van Gennep). Her current research focuses on technology and the history of noise.

HANS-JOACHIM BRAUN has been a professor of modern social, economic and technological history at the Universitaet der Bundeswehr Hamburg since 1982. His publications focus on technology transfer, failed innovations, production technology, engineering sciences, the social history of engineers and technology and music relationships. He is Secretary-General of the International Committee for the History of Technology (ICOHTEC), chairman of the scientific council of the Georg-Agricola Society and editor of *Studien zur Technik-, Wirtschafts- und Sozialgeschichte*.

MARTHA BRECH is a lecturer in systematic musicology at the Technical University of Berlin. Her background is in musicology, sound engineering, ethnomusicology and German studies. She received her M.A. degree in ethnomusicology in 1984 and her Ph.D. with an analysis of electro-acoustic music by means of sonograms in 1993 (Frankfurt am Main: Peter Lang, 1994). Her publications are on musicology and twentieth-century music.

HUGH DAVIES is a freelance composer, performer, instrument inventor and musicologist, specialising in twentieth-century musical instruments and electronic music. He was the founder-director and a research consultant for the Electronic Music

Studio, University of London Goldsmiths' College, as well as an external consultant for electronic musical instruments at the Gemeentenmuseum, the Hague. His writings include *International Electronic Music Catalogue* (1968, compiler), contributions to eight dictionaries, including *The New Grove Dictionary of Musical Instruments* (1984, 305 entries) and chapters in numerous books and exhibition catalogues. His compositions include chamber music, electronic music, works for his invented instruments and environmental sound documentation. Since 1999 he has been a part-time researcher in Sonic Art at the Centre for Electronic Arts, Middlesex University, London.

BERND ENDERS holds a chair at the University of Osnabrueck of systematic musicology with special emphasis on music and technology. He also co-ordinates the biannual KlangArt New Music Technology convention and is author of *Lexikon Musikelektronik*, 3rd ed. (Zuerich, 1997) and is editor of *Neue Musiktechnologie*, Vol. 1 (Mainz, 1993) and *Neue Musiktechnologie*, Vol. 2 (Mainz, 1996).

GEOFFREY HINDLEY lectures in European and British civilization at the University of Le Havre and is President of the Society for the History of Medieval Technology and Science, Oxford and London. He graduated in history from Oxford (1958) and is a writer, editor and lecturer in history and the history of music. His numerous books include: *The Shaping of Europe, Castles of Europe, The Medieval Establishment, Medieval Warfare, Saladin: A Biography, England in the Age of Caxton* and *The Book of Magna Charta*. His *Larousse Encylopedia of Music* (general editor, 1997) is in its tenth printing and his other publications in music include *Musical Instruments: A History* and 'Le Clavier – musique et technologie' (in *Cahiers de la musique ancienne*, Nice).

JÜRGEN HOCKER has been chairman of the Society for Automatic Musical Instruments since 1980. In 1986 he bought and restored an Ampico-Boesendorfer-player piano which he used in performing numerous concerts worldwide, particularly the player piano works of Conlon Nancarrow. At the *Musik Triennale* Cologne in 1997 he performed Nancarrow's complete works for player piano. He has cooperated with György Ligeti since 1990.

MARK KATZ received his Ph.D. in musicology from the University of Michigan and now teaches at the Peabody Conservatory at the Johns Hopkins University. His other work on the influence of recording may be seen in 'Making America More Musical through the Phonograph, 1900-1930' in the winter 1998 issue of *American Music*. He is currently working on *The Phonograph Effect*, a broad study of recording's impact on modern musical life, which will be published by the University of California Press.

TATSUYA KOBAYASHI is a professor of technology and culture at Chukyo University in Nagoya, Japan. His main field of research is technology transfer and innovation in the formative years of modern Japan. One of his best-known books is *Technology Transfer – A Historical Study. America and Japan* (Bunshindo, 1981). He is a member of the Executive Committee of ICOHTEC.

JAMES P. KRAFT is an associate professor at the University of Hawaii at Manoa. He received his Ph.D. from the University of Southern California in 1990, where he specialized in business and labor history. He is the author of *Stage to Studio: Musicians and the Sound Revolution, 1890-1950* (Baltimore: Johns Hopkins University Press, 1996).

ALEXANDER B. MAGOUN received his M.A. in history at the University of East Anglia in 1983. During the 1980s he was a disc jockey at WRTC-FM in Hartford, CT, and continues to deejay independently. He is writing his dissertation in the history of technology at the University of Maryland at College Park, *Shaping the Sound of Music at RCA Victor, 1929-1958*, and is currently curator of the Sarnoff Corporation, Princeton, NJ.

REBECCA MCSWAIN is a research associate in anthropology at the University of Colorado at Boulder. Since receiving a Ph.D. in anthropology from the University of Arizona in 1989, she has pursued research interests in ancient Maya lithic studies and in the realm of modern material culture. She has done field work in arenas as disparate as the lowland Maya area of Belize and regional vintage-guitar shows across the US, and she has written and lectured on both of these areas of interest. Her most recent publication is 'The Power of the Electric Guitar' (*Popular Music and Society*, 1995). She is currently at work on a book entitled *The Blue Guitar and the American Dream*, an assessment of the cultural role and significance of the electric guitar.

ANDRE MILLARD is a professor of history and director of American Studies at the University of Alabama at Birmingham. He teaches history of technology in one department and popular culture studies in the other. These two fields of interest have been combined in his latest book, *America on Record: A History of Recorded Sound* (Cambridge University Press, 1995), in which he considers some of the cultural impact of systems of sound recording. His other publications have dealt with the career of Thomas Edison and the work of his West Orange Laboratory. Currently Millard leads the Birmingham Music History Project, an oral history of music and musicians in Alabama.